Rising Force

Rising Force

The Magic of Magnetic Levitation

James D. Livingston

Harvard University Press • Cambridge, Massachusetts • London, England • 2011

Library of Congress Cataloging-in-Publication Data

Livingston, James D., 1930–
Rising force : the magic of magnetic levitation / James D. Livingston.
p. cm.
Includes bibliographical references and index.
ISBN 978-0-674-05535-3 (alk. paper)
1. Magnetic fields.   2. Magnetic suspension.   I. Title.

QC754.2.M3L58   2011
538—dc22   2010047290

Contents

My earlier popular-science book, *Driving Force: The Natural Magic of Magnets* (Harvard University Press, 1996), covered a wide variety of topics in the history, science, and technology of magnets. Most of the history and basic science presented there remains accurate, but many of the technology areas have advanced considerably since 1996. Technology is a moving target. In particular, magnetic levitation, that is, the use of magnetic forces to combat gravity and friction, has recently advanced in a wide variety of technology areas.

The world's first commercial maglev train line was constructed in China to link the Shanghai airport to the city center, and, since 2003, millions of passengers have already experienced the excitement of traveling at speeds up to 250 miles per hour. A less visible but much wider application of magnetic levitation is in magnetic bearings. Here maglev applications have greatly increased in recent years and include their use in artificial hearts, energy storage, wind turbines, integrated-circuit manufacture, and ultracentrifuges to enrich uranium. (The recent assembly of large cascades of such centrifuges in Iran has become a major issue of international politics.)

"Flying frogs," the levitation of living frogs in high magnetic fields, drew much public attention in 1997 and stimulated greatly increased attention to the general area of diamagnetic levitation— levitation of matter repelled by magnetic fields. Superconductors, the ultimate diamagnetic materials, have seen increased use in maglev devices and demonstrations, including the levitation of a Japanese sumo wrestler and, of considerably more scientific importance, a half-ton superconducting ring for fusion research.

The Levitron, a very popular toy in which a spinning magnet achieves stable levitation through gyroscopic action, became widely distributed in the late 1990s, and improved models, with higher levitation heights, have appeared in recent years. Many aesthetically striking items employing sophisticated electrical circuitry to achieve desktop levitation of globes and other magnetic objects have been developed and have even appeared in the form of a floating sculpture in an art museum.

All these and other recent developments have convinced me that it would be timely to write a full-length popular-science book focused on one particular aspect of "the natural magic of magnets"—the topic of magnetic levitation. To quote from *Driving Force,* "Does gravity get you down? Magnets can lift you up! Does friction slow you down? Magnets can speed you up! Fighting the forces of gravity and friction is one of the things that magnets do best."

I should perhaps note here that some authors reserve the term *magnetic levitation* for cases where the magnetic antigravity force is *repulsive,* delivered from below, and prefer the term *magnetic suspension* for cases where the magnetic antigravity force is *attractive,* delivered from above. But most regard the words "suspension" and "levitation" as virtual synonyms. In this book, I will use the term *magnetic levitation* for magnetic antigravity forces of all types, in part because I much prefer the shortened form "maglev" to "magsusp."

We'll start in Chapter 1 with examples of humankind's longtime fascination with levitation from the worlds of literature, films, television, theater, magic, and religion, and then briefly describe various physical but *nonmagnetic* means that humankind has developed to combat gravity. Since magnetic levitation involves the use of magnetic forces to combat gravitational forces, we'll review in Chapter 2 the basics of those two competing forces, and in Chapter 3 the fundamentals of maglev, including the central problems of force balance and stability. Later chapters describe the various types of magnetic levitation that have been developed and their applications, finally reaching in Chapters 11 and 12 the topic of maglev trains, one of the most dramatic examples of magnets "fighting the forces of gravity and friction. . . one of the things that magnets do best."

Rising Force

Levity vs. Gravity

Fictional and Illusional Levitation

A few years ago I was in Manhattan on a short business trip, staying in a hotel in the theater district. I had a free evening, decided to see a musical, and chose a performance of *Mary Poppins,* hoping to relive some of the fun of the earlier movie version starring Julie Andrews and Dick Van Dyke. I enjoyed the show, but the most memorable and magical moment came at the very end. After Mary Poppins had said her goodbyes to the Banks family, she opened her umbrella and flew away. And this time she didn't just fly across the stage and into the wings, as she had done earlier in the show. This time she held up her umbrella and casually flew off the stage, up and high over the theater audience, up and high over the balcony, and out of sight. It was a wonderful effect, and its magical impression on me was enhanced by the wonder expressed in the face of the 6-year-old girl who happened to be sitting next to me. There's something about seemingly conquering the ubiquitous force of gravity that is especially magical to most of us.

I was well aware that the flight of Mary Poppins above my head was not really magic, not really a miracle. The rational part of my brain knew that she was supported by a harness attached to an array of many fine wires, wires fine enough that they would not be visible to us in the limited lighting in the heights of the theater above our heads. Nevertheless, I enjoyed the feeling of magic. Suspension of disbelief can be an important part of enjoying the theater. I'm not sure of the thinking of the 6-year-old girl who sat next to me. The world remains full of mysteries at that age, and I suspect she was not then thinking about harnesses and arrays of wires. During intermission, her mother had bought her a replica of the magical umbrella that seemed to be an important part of the flying ability of Mary Poppins, and probably in the following weeks, the girl often opened her umbrella and pretended to fly. Pretense and imagination are important parts of the gifts of childhood, a gift we adults often lose. If the girl took any time at all to wonder about how Mary Poppins was able to fly, she perhaps explained it to herself by the effects of a special wind on the unfurled umbrella. After all, the flying of kites was an important part of the play, and isn't it wind that makes kites fly?

In the book and stage versions of *Peter Pan,* the ability to fly was explained by Peter Pan as requiring only fairy dust and happy thoughts. (In Disney's movie version, "fairy dust" was converted to "pixie dust," probably because pixies were considered more politically correct.) In the TV series *The Flying Nun,* Sally Field wore an unusual winglike headdress that perhaps allowed a magical wind to levitate and propel her through the air. Disney's Aladdin used a flying carpet, and Harry Potter and other students at Hogwarts flew on broomsticks, as did the scary Wicked Witch of the West in *The Wizard of Oz* (who doesn't seem quite so scary in "Wicked," where she sings "Defying Gravity."). From the Middle Ages on, broomsticks have been portrayed as the preferred flying mechanisms for witches, the flying power of the broomsticks often enhanced by application of magical ointments. (Most modern attempts to explain the origin of this long-held superstition have a sexual component, with the broomstick a phallic symbol, a notion not stressed in the Harry Potter books.) In the *Star Wars* films, Yoda and other Jedi Masters are able to levitate objects with the use of "The Force," while Chinese martial arts films like *Crouching Tiger, Hidden Dragon* feature warriors battling in graceful gravity-defying displays with no attempted explanation of the source of their abilities. To fly around the world on Christmas Eve, Santa Claus requires a team of magical reindeer, but

Superman and many other superheroes in the world of fiction somehow levitate and fly on their own power, with no reindeer, broomsticks, umbrellas, winged hats, flying carpets, pixie dust, or even happy
thoughts required. The ability to counteract the downward pull of the
earth clearly exerts a powerful attraction to our imagination.

Demonstrations of levitation have also long been a popular tool
for magicians to display their powers. One famous illusion was the
"ethereal suspension" introduced by French magician Jean Eugene
Robert-Houdin in 1847. As Robert-Houdin himself described it in
his memoirs,

> I placed three stools upon a wooden bench. My son stepped on the
> middle one; I had him extend his arms so I could support him with
> two canes, each of which rested on a stool . . . I removed the stool so
> the child was supported only by the two canes. This strange balanc-
> ing already evoked great surprise among the spectators. It grew even
> more when they saw me remove one of the two canes and the stool
> that supported it, and it reached its peak when, after having raised my
> child to a horizontal position using my little finger, I left him sleeping
> in space, and to defy the laws of gravity, I also removed the feet of the
> bench at the base of this impossible edifice.

His son was of course supported by the other "cane," which was
actually a strong iron bar connected at the bottom to a firm base and
at the top to a mechanical harness worn by the son but hidden underneath his clothes. The harness and cane were joined with a complex
but invisible linkage that allowed the magician to swing his son
from vertical to horizontal. The illusion was featured on a 1971 post-
age stamp of France honoring the centennial of Robert-Houdin's
death. It shows his son horizontal in the air, supported under his
arm by a thin cane standing on a stool mounted on a seemingly unsupported bench. (Both the son and the bench were cantilevered.)

This trick has evolved to the current day in various forms, one
of which is the "broomstick suspension" in which Robert-Houdin's

two canes are replaced by two broomsticks with their bristles facing up. This version even appeared in one episode of TV's *I Love Lucy*, with Orson Welles as the magician and Lucille Ball as "Princess Lu Cy" seemingly supported only underneath one arm by the bristles of a broomstick (containing a concealed iron bar). Harry Houdini later honored the memory of Robert-Houdin by borrowing part of his name, but in *The Unmasking of Robert-Houdin,* wrote that Robert-Houdin stole the "ethereal suspension" and others of his famous tricks from earlier magicians.

An earlier and simpler illusion, involving no body harness and no complex linkage to the vertical rod, consisted simply of a metal platform cantilevered a few feet above the ground, supported on one end by a firmly anchored vertical iron rod. The levitated person wears loose clothing to conceal the platform, and the clothing often hangs down to conceal the supporting rod. Where full concealment of the rod is not possible, he simply puts one hand on the upper end of the rod/cane "to keep his balance." This simple illusion is considerably older than Robert-Houdin's "ethereal suspension," but it created a sensation at an outdoor exhibition in South India in 1936, largely because photographs were taken and published in the *Illustrated London News.* The pictures, which can be seen today on the Web, show Yogi Subbayah Pullavar seemingly levitating a few feet above the ground in front of a crowd of 150 witnesses. Before the demonstration, Pullavar, platform, and rod support were inside a tent so that the assembly of the illusion could not be observed. Once it was ready, the tent flaps were opened for several minutes to amaze the audience with the levitating Yogi, and then closed again to allow the illusion to be disassembled. Recently a Dutch magician with the stage name of Ramana has revived this "Indian magic," performed the illusion on television, and even demonstrated it in Times Square.

Of course, more effective are levitation illusions in which there is no visible supporting cane or rod or broomstick, illusions in which the levitated body appears to be completely unsupported, simply floating in air. John Nevil Maskelyne, an English magician of the nineteenth century, is usually credited with the invention of the first effective illusion of this kind. Here the magician's assistant could actually be seen to rise into the air (aided, of course, by magical hand gestures of the magician). It involved an assembly of fine wires to lift the platform on which the magician's assistant reclined. The structure was cantilevered and supported by an iron pillar that stood about a foot behind the platform and could slide up and down through the stage. The curvature of the rod connecting the platform to the pillar allowed Maskelyne to stand in front of the pillar and obscure it from the audience. The audience's view of the pillar was often also partly blocked by the assistant's loose clothing that hung below the platform as he or she was lifted. The most important component of the illusion was the "gooseneck" in the rod of the support structure, which allowed Maskelyne to pass a circular hoop over the levitated body "not once, but twice" to convince skeptics in the audience that no wires or pillars were involved. This was possible because although the hoop did indeed pass over the levitated body from head to toe, it avoided the wires and the pillar. After the first pass, the hoop was still linked into the gooseneck of the support structure, but by pulling it back through the gooseneck and passing it over the levitated body again, the hoop became free. So passing the hoop over the body "not once, but twice" was an absolutely necessary part of the illusion.

In *Levitation: Physics and Psychology in the Service of Deception*, Jim Ottaviani and Janine Johnston diagram in detail the Maskelyne "levi" trick, report how it was stolen by American magician Harry Kellar (Figure 1), and show how Kellar improved it, replacing the pillar with a second array of fine wires invisible to the audience. With no pillar to hide, the magician could now walk completely around the levitated body, usually the body of an attractive young woman appearing in Oriental clothing and introduced as Princess Karnak "from the mysterious East." (It was helpful if the "princess" was not only attractive and young but also not too heavy, thereby limiting the stress on the wires.) As with most illusions presented by magicians, the levitation was introduced and accompanied by an extensive

Figure 1. 1894 poster advertising famed levitation trick of magician Harry Kellar.

colorful background story and mysterious incantations. Kellar later sold this and others of his successful tricks to his successor Harold Thurston, who for many years was America's most popular magician. Thurston often invited some people from the audience to come to the stage to convince everyone that Princess Karnak "actually floats in space without any support." Once on the stage, the men and women from the audience could of course see all the wires but, being told that they were now "part of the story," they usually did not publicly reveal the magician's secrets. Why spoil a good show for others? As Ottaviani and Johnston note at the end of their book, "that's magic."

One of the most impressive demonstrations of levitation by magicians today is that of David Copperfield. I first saw it on television, and it can be seen today online via the wonders of the Worldwide Web. In this performance, instead of levitating a recumbent "princess," Copperfield gracefully flies himself, gliding through the air and even turning somersaults in midair. Like Maskelyne did in the nineteenth century, Copperfield dispels your suspicions of supporting wires by having assistants seemingly pass circular hoops around his body. He later descends into a large transparent box, assistants put a cover on top of the box, and he can even levitate while in the box! He sometimes adds a sequence in which he invites a woman in the audience to join him, and he flies with her in his arms, as Superman did with Lois Lane.

The Web also includes many sites explaining Copperfield's levitation, which involves—you guessed it—an array of fine wires attached to a harness around his waist. This particular supporting system was created and patented by John Gaughan, a leading manufacturer of equipment for magicians. It is more sophisticated than earlier levitation devices and allows a wide variety of motions while the performer is in the air. As for those circular hoops, by using two assistants manipulating the hoops very rapidly and in synchronism, the illusion is created of hoops passing around Copperfield's body even though they avoid the wires suspending him. And when he enters the transparent box, the top of the box is in sections that allow the wires into the box and permit Copperfield limited levitation even inside the box. But you might prefer just watching the magician's wonderful performance without thinking of all the complex paraphernalia involved.

More down-to-earth levitation illusions include the "street magic" of David Blaine and others that can be performed anywhere without the need of overhead wires. Here, however, the apparent levitation lifts the performer only a few inches above the ground and lasts only a few seconds. One such illusion requires no special equipment at all, just skill and lots of distracting patter and arm motions. Here the performer retreats a few feet from a small group of adult observers, turns his back to them, and stands at an angle to them. He then slowly rises on tiptoe on the foot away from the onlookers, while lifting his other foot so that his two heels stay together, giving the appearance from behind that both feet are being lifted off the ground. The foot nearer the audience blocks their view of the tiptoe position of the other foot. The "levitation" must be brief, and there should be no children in the audience because they are likely to bend over quickly to look underneath the performer's front foot and detect the true source of the "levitation."

There are several other techniques for "street magic" levitation, some of which involve movable supporting pillars hidden inside the performer's trousers, but it of course remains important that the supporting device, be it a pillar or the front of a foot, be blocked from view. So this form of levitation must be limited in both height and duration. Also very popular with magicians and their audiences are tricks in which the levitated object is not the performer but smaller and lighter objects such as playing cards. In such cases, you are right to suspect the presence of some very thin unseen strings. Even light objects like playing cards need help to resist the relentless pull of gravity.

Wingardium Leviosa!

Levitation plays an important role in the fantasy world of the Harry Potter books and films, and not only with flying broomsticks. In the first book of the series, the young wizard students at Hogwarts Academy are pleased when in their Charms class, Professor Flitwick announced that they would be finally learning the levitation spell. On each student's desk was placed a large feather to levitate. (After all, these were only young wizards trying their first levitation, and after they first learned to levitate light objects like feathers, they could later progress to heavier things.) The professor explained that, like most spells, it would require both proper use of their magic wands (the correct "swish and flick") and saying the magic words properly. For the levitation spell, the magical incantation was "Wingardium Leviosa!"

Harry's friends Ron Weasley and Hermione Granger were sitting together in the front of the class and Ron tried first. After a wide, sweeping movement of the wand, he pointed it at the feather in front of him and said, "Wingardium Leviosa!" But the feather did not budge. Hermione then explained to Ron that he was saying it wrong, emphasizing the wrong syllables. Hermione then picked up her wand, gave it the proper "swish and flick," and said "Wingardium Leviosa!" with the proper pronunciation. The feather rose from the desk and hovered above their heads. "Oh, well done!" cried Professor Flitwick, "Everyone see here, Miss Granger's done it!" Neither Ron nor the other students in the class were all that pleased that Hermione had mastered the levitation spell first.

Later in the story, however, Hermione and Harry were in serious danger from a giant troll carrying a huge club. Fortunately, by now Ron had learned the levitation spell. After a wave of his wand and a well-pronounced "Wingardium Leviosa!" the troll's club rose in the air and then dropped down on the troll's head, knocking it out and thereby saving Hermione and Harry. In a later episode, Harry also used the levitation spell to escape a difficult situation. Apparently, although a girl was the quickest to learn, boys can eventually catch up—at least in the fantasy world of Harry Potter.

In his book *The Science of Harry Potter: How Magic Really Works,* British author Roger Highfield uses the wonder in the Harry Potter stories as an excuse to introduce some of the "magic" in real-world science. With regard to various examples of levitation in Harry Potter, Highfield seems to feel that some form of magnetic levitation is probably at work, which he introduces in a section entitled "Magnets, the Levitron, and Levitating Frogs" (topics we discuss in Chapters 4 and 6). But he makes no attempt to give a scientific explanation of why the levitation in Harry Potter's world seems to depend on the proper pronunciation of the magic words.

In Harry Potter's world, the nonwizard humans are called Muggles. In our world, however, the real wizards are the Muggle inventors, scientists, and engineers who use their knowledge and creativity, not magic wands and magic words, to produce all the magic of modern technology, including the magic of magnetic levitation.

Mystical Levitation

David Copperfield introduces his levitation show with references to the common dream of flying. I've had that dream, and you probably have too. In his famous *The Interpretation of Dreams,* Sigmund Freud confesses to be mystified by flying dreams. Since such dreams are generally pleasurable, Freud suggested that they might just be a remnant of childhood fun with swings or seesaws or being lifted and swung in the air by playful parents or uncles. (He of course also suggests a possible sexual interpretation, as is his wont.) Other writers suggest that flying dreams may represent a desire for freedom from the limitations of everyday life, or perhaps a wish for social elevation. Whatever the meaning of flying dreams, I suspect that even a few saints have had them—or at least had some mystical experiences akin to dreaming.

Copertino, once spelled Cupertino, is a small town in southern Italy, in the "heel" of the boot-shaped Italian peninsula. The town is

most famous for the birth there in 1603, in a stable, of Giuseppe
Maria Desa, known in the West today as Saint Joseph of Cupertino.
Giuseppe grew up to be very devout. He became a friar and often was
gripped in intense religious "raptures" or "ecstasies," trances in which
he had miraculous visions. Stories began to circulate that while he
was in these trances, he often levitated several feet above the ground.
Witnesses reported that over the years he levitated dozens of times,
sometimes to considerable heights, and once even levitated in the
presence of the Pope. After Giuseppe's death in 1663, stories of his
levitations were part of the reports issued for his beatification, and
he was canonized in 1767. Several paintings of the saint show him
floating in mid-air. San Giuseppe de Copertino, or Saint Joseph of
Cupertino, the "Flying Friar," is now, most appropriately, the patron
saint of pilots and air travelers.

Perhaps the best known and most loved of the many other Catho-
lic saints who have been reported to levitate is Saint Teresa of Avila.
In her case, we can read her own accounts of her levitations in her
autobiography, written in 1562 for her confessors. In it, she often
reports seeing visions of devils and angels and even the presence of
Jesus himself, and describes her frequent raptures, transports, and
"flights of the spirit." "My soul has been carried away," she wrote,
"and usually my head as well, without my being able to prevent it;
and sometimes it has affected my whole body, which has been lifted
from the ground." She was not pleased with her levitations. She asked
her fellow nuns not to speak of them, and on one occasion, asked them
to hold her down. She beseeched the Lord "to grant me no more fa-
vors if they must have outward and visible signs." She was apparently
aware that some might think that her levitations were only dreams.
"One does not lose consciousness," she argued, "At least I myself was
sufficiently aware that I was being lifted."

Considering the times in which she lived, Teresa was certainly
an outstanding woman. In addition to her autobiography, Teresa
wrote several other books and spent many years traveling throughout Spain founding Carmelite convents. She was so revered during

her life that she was canonized only forty years after her death, and later became one of the patron saints of Spain. The desire to claim a piece of divinity was then very powerful. Teresa's body was exhumed and dissected, and her relics became highly prized and distributed to various holy sites throughout Europe. General Francisco Franco reportedly kept her left hand by his bedside throughout his life. Her home city of Avila, with a restored center that is a World Heritage Site, retained only one of her fingers. When my wife and I visited Avila several years ago, we were most charmed by the ancient stone walls that surround the central city, but our tour guide clearly felt that the most important wonder of Avila was Saint Teresa's finger.

In the Old Testament, Ezekial reported being visited by a spirit who "lifted me up between the earth and the heaven." Gnostic books tell of the levitation of Simon Magus above the apostle Peter, and Islamic sources tell of King Solomon traveling through the air, albeit with the aid of a magic carpet. There are tales of levitating Japanese ninjas and Indian yogis and fakirs (some, like Subbayah Pullavar, may also have been fakers). Tibet has legends of levitating lamas, and the Buddha himself reportedly had the power, although it is said that he was reluctant to levitate in the presence of others, since he disapproved of the public demonstration of miracles. In modern times, practitioners of Transcendental Meditation practice "yogic flying"—essentially a short hop with the legs crossed in the lotus position—which is claimed to be the first step toward full levitation. (But no transcendental meditators yet seem to have reached the second step.) Many people have described the sensation of flying during "out of body experiences" (OBEs), such "flights of fancy" sometimes enabled by hallucinogenic drugs. Amazon.com offers several books on self-levitation, including a "how-to manual."

Humans can fly in their dreams and mystical raptures, and from ancient times to today, have developed countless stories of gods and heroes (and sometimes villains) capable of levitation. The desire to overcome, at least to a degree, the force of earth's gravity seems to have been with us for a very long time, probably since the very first humans wondered at—and envied—the wonder-full flying abilities of birds. Fortunately, we are not limited to fictional, illusional, and mystical means of levitation. There are several real physical methods to produce upward forces and achieve levitation.

Physical Levitation without Magnets

It was in Paris in 1783 that humans first achieved free flight. Brothers Joseph and Etienne Montgolfier had been inspired by observing smoke rise up the chimney, and explored the possibility that the "levity" of smoke could be exploited to lift other objects. They started with a small silk bag that they held over a fire, and the bag rose to the ceiling. They then advanced to much larger bags and to public demonstrations, including one before Louis XVI at Versailles in which a sheep, a rooster, and a duck were the passengers. The balloon floated for several minutes and landed about two miles from the launch site. The animals looked none the worse from their trip, so the brothers were encouraged to try human passengers. After trials with humans in baskets attached to tethered balloons, a Montgolfier balloon carrying two French noblemen as volunteers was released in the Bois de Boulogne on November 21, 1783. This first manned free flight lasted about 23 minutes, rose to an altitude of over 3,000 feet, and landed safely several miles away. Humankind was finally freed from the confines of the earth's surface.

It took a while for the Montgolfier brothers to learn that it was not the smoke itself that conveyed levity to the balloon, but the heated air that was less dense than the cooler air outside the balloon. The balloon ascension did not actually overcome gravity—it used it. Gravity pulled downward on a given volume of the surrounding unheated air with a greater force than it pulled downward on the lighter heated air within the balloon, so the surrounding unheated air fell and the balloon rose. Earlier, in 1783, other Frenchmen had used another approach to produce a gas lighter than air, filling a balloon with the recently isolated gas, hydrogen, to successfully launch an unmanned balloon. Hydrogen balloons had distinct advantages over hot-air balloons, which required considerable effort during flight to keep alive the fire that heated the air. However, hydrogen also has a distinct disadvantage as a lifting gas for passenger flight—it burns. It was not until 1937, with the fiery end of the German airship *Hindenburg*, that hydrogen was finally abandoned as a common lifting gas. Today's lighter-than-air ships use helium, a nonflammable, chemically inert gas first identified in the solar spectrum in 1868 (named from *helios,* Greek for the sun) and not discovered as a significant component of natural gas until 1903. At ordinary temperatures, although helium is about twice as dense as hydrogen, it is still about seven times lighter than air, so it can provide plenty of lift.

The phenomenon of buoyancy that lifted the Montgolfier hot-air balloon and today lifts the Goodyear blimp and other lighter-than-air ships, as well as many party balloons, is the same thing that enables me to do something I enjoy each summer—floating on my back in ocean water. I can do that because the salt water is a bit denser (i.e., weighs a bit more for a given volume) than my body. So the earth's gravity pulls down harder on the salt water than it does on me, and I, a lighter-than-salt-water body, can float. As reported by Archimedes in the third century B.C.E., a body immersed in a fluid experiences a buoyancy force equal to the weight of the displaced fluid. (The familiar story that this discovery led Archimedes to jump out of his bathtub and run naked into the streets crying "Eureka" is highly questionable.) The buoyancy force can also be viewed as the difference between the fluid pressure exerted on the bottom surface of the immersed body (pushing it up) and that exerted on its top surface (pushing it down). In any fluid, be it liquid or gas, pressure increases with depth.

Here I probably should mention that there is a magnetic form of buoyancy commonly called the *magneto-Archimedes effect.* For this you need a magnetic liquid or *ferrofluid,* which consists of extremely fine particles of a magnetic material, commonly an iron oxide, suspended in a water-based or oil-based liquid. Pour the ferrofluid into a container, immerse a small chunk of copper or other nonmagnetic material in the ferrofluid, and put a strong permanent magnet under the container. The ferrofluid is attracted strongly downward toward the magnet, it flows underneath the copper, and the copper rises. (This effect occurs even in weaker magnetic liquids, as long as the immersed object that you want to lift is less magnetic than the liquid.) Some call this a form of maglev, but for this book, I'll reserve the term *levitation* for cases where an *upward* magnetic force opposes the downward force of gravity, and the lifted object is contact-free or nearly contact-free and not immersed in liquid.

Let's return to the more relaxing topic of me floating on my back in ocean water. For those who like numbers, I note that the density of fresh water is 1 gram per cubic centimeter (g/cc) or 62.4 pounds per cubic foot, while the density of most ocean water is a bit higher, about 1.03 g/cc. Ocean water is about 3% denser than fresh water because, by weight, it holds about 3% salt. (The water in special places like Utah's Great Salt Lake or Israel's Dead Sea holds much more salt and can have densities as high as 1.2 g/cc or even greater.) The density of the human body is close to that of water but is variable. Your bones and muscles are denser than fresh water, but your fat is less dense. With more fat, you float more easily. And you can change your average density by varying the amount of air in your lungs. If I blow the air out of my lungs, I rapidly sink to the bottom of a swimming pool—I'm rather skinny. But not in the ocean—that 3% extra density of the water, provided by the salt, makes it easy for me to float.

In the above paragraph, I used two different units for mass or weight—grams and pounds—and two different units for distance—centimeters and feet. (To be more precise, I used two different units for volume—cubic centimeters and cubic feet.) In my work as a scientist, I usually use the metric system, which includes describing masses in grams and distances in meters, but we Americans as a society have been very reluctant to give up the "English system" of ounces and pounds and inches, feet, and yards, even though the English themselves abandoned it long ago. I'm an American and a scientist (a Scientific American?), so I feel reasonably comfortable with both sets of units, but some readers of this book may be either nonscientists or non-Americans, or both, and I want you *all* to feel as comfortable as possible. So when I need to use numbers and units to describe things quantitatively in this book, I'll usually provide a translation from one set of units to the other, as I did above. In case I forget, you can refer back here for the basics: 1 pound = 454 grams = 0.454 kilogram, and 2.54 cm = 1 inch = 1/12 foot.

Human flight started with the Montgolfier brothers and buoyancy. The next major step came 120 years later and was also made by two brothers—Wilbur and Orville Wright. In December 1903, on the sand dunes of Kitty Hawk, North Carolina, they demonstrated that even an object heavier than air could be encouraged to fly and carry a human aloft with the help of wings, a headwind, and engine-driven propellers. They started with kites and gliders, and their first successful biplane was essentially a powered glider. Today's huge passenger airplanes, like the Wrights' *Flyer* of 1903, get their levitation or lift mostly from the difference between the upward air pressure exerted on the lower surface of the wings and the downward air pressure on their upper surface. In the Montgolfier balloon, this pressure difference between bottom and top resulted directly from the force of gravity and the resulting increase of air pressure with depth. In the Wright biplane and modern airplanes, it results mostly from the plane's motion through the air, from the "wind beneath my wings." Lifting forces generated by propellers in helicopters and by fans in hovercrafts (air-cushion vehicles) also result from differentials in air pressure. On a much smaller scale, an upward stream of air, for example, from a hair dryer, can levitate a ping-pong ball, a process that has been called *aerodynamic levitation* when used in a laboratory for research purposes.

You can lift off the earth yourself with a *downward* stream of gases if you have the nerve to try a rocket belt of the kind seen in the James Bond 1965 movie *Thunderball* and featured much earlier in the Flash Gordon comics. Here the downward momentum of the escaping gases yields a reverse upward thrust that can lift a human wearing the belt, but only for about 20 seconds before the fuel runs out. NASA's rockets lifting shuttles into orbit or astronauts toward the moon provide upward thrust much longer, since they can carry a lot more fuel than James Bond could carry on his back.

A less familiar way to produce antigravity forces is *acoustic levitation*—levitation by sound. Sound waves in air are waves of varying air pressure, usually generated by vibrations of surfaces in contact with the air, like the diaphragms of the speakers in radios, telephones, and other electrical devices (or our own larynxes). Scientists have used sound waves to demonstrate acoustic levitation of various solid objects, even including living things like beetles, spiders, ants, and ladybugs. The sound waves are generated utilizing transducers—devices that convert electrical energy into sound energy, usually piezoelectrics, materials that change dimensions under electrical stimulation. Those changes in dimension produce vibratory surface motions that generate vibrations in the air, that is, sound. Most sound waves are traveling waves in which the maxima and minima of air pressure (regions of compression and rarefaction) move outward from the source, for example, the TV speaker. Optimum acoustic levitation requires a sound reflector and interference between incident and reflected waves to create "standing waves" of sound between the emitting and reflecting surfaces, waves in which the maxima and minima of air pressure remain fixed in space. Although acoustic levitation of beetles is amusing (perhaps not to the beetles), the technique of acoustic levitation also has practical applications in science and engineering, such as the levitation of molten materials that are highly active chemically and would react if in contact with most possible containers. Acoustic levitation is one laboratory technique used for "containerless processing" of reactive materials, as well as for other experiments in which study of a material without the influence of a contacting surface is of interest. More generally, simulated weightlessness is of special interest, of course, to NASA. Many experiments done by NASA and other laboratories using acoustic levitation employ "sound" waves of frequencies higher than we can hear, that is, higher than 20,000 cycles per second. The technique is then called *ultrasonic levitation.*

Except for the rocket belt, all the physical techniques that produce levitation mentioned so far require the presence of air or another gas to produce the upward force on the levitated object. However, several levitation techniques can operate in a vacuum without the use of rockets. One such technique is *optical levitation,* in which the upward force on the levitated object is provided by light, usually in the form of a high-intensity laser beam. Instead of air pressure, the object is levitated by "radiation pressure" produced by the particles of light, called photons, which carry momentum and transfer it to the levitated object. With enough laser intensity, stable suspension of small solid particles or liquid droplets has been achieved.

Another physical levitation method that can operate in vacuum is *electrostatic levitation.* NASA's Marshall Space Flight Center in Huntsville, Alabama, offers an electrostatic levitator for researchers interested in containerless processing. It has two horizontal copper plates with a large voltage difference between them, producing a large vertical electric field between the plates that yields an upward force on a charged object. With the use of side electrodes, complex electronics, and techniques to maintain charge on the object, stable levitation can be achieved. It's a highly sophisticated use of the electrostatic force, which is most familiar to us via the "static cling" of clothing removed from the dryer.

Electrostatic levitation is interesting, but more complicated to achieve and less widely useful than its close cousin, the main subject of this book—*magnetic levitation,* or, in its popular abbreviated form, *maglev.* To appreciate the challenges of maglev, we should first review the basics of gravitational and magnetic forces, which we do in the next chapter.

Gravitational and Magnetic Forces

Forces of the Universe

Theoretical physicists tell us that there are only four basic forces of the universe: gravitation, the electromagnetic force, and the strong and weak nuclear forces. In this book on maglev, we can ignore the two nuclear forces and concentrate on the competition between gravitational and magnetic forces. But what about the force we're most familiar with, the *force of touch*—the force the bat exerts on the baseball, the antigravity force our chair exerts on the seat of our pants, the force of air pressure that, as we discussed in the previous chapter, lifts helium balloons and 747s off the ground? Physicists think of such things on the atomic level and tell us that the force one object exerts on another through touch is basically the repulsion at very small separations between electrons in one object and electrons in the other. So the force of touch is just one aspect of the electromagnetic force. So let's forget about touch for now and concentrate on *magnetic levitation,* that is, the competition between upward magnetic forces and downward gravitational forces—two forces that operate at a distance *without touch*. First we should review a few basics about gravitation. We should know the enemy.

The gravitational force between two masses, unlike electric and magnetic forces, is *always attractive.* According to Isaac Newton, the gravitational attractive force between two masses varies as the product of the two masses—mass 1 times mass 2. If both of the two masses are small, the gravitational force is tiny. However, we live on the surface of a huge mass—the earth—that exerts a very large and very noticeable downward force upon us. That's the force we're trying to oppose with maglev. Newton also tells us that the gravitational force between two masses varies as the *inverse square* of the distance of separation between their centers. That means that if you double the distance of separation between two centers of mass, the force decreases by a factor of four—two squared ($2^2 = 2 \times 2$). If you triple the distance of separation, the force decreases by a factor of nine—three squared ($3^2 = 3 \times 3$). That's the famous *inverse-square law*.

The gravitational force we're most interested in is the force exerted by the earth on us and other objects near the surface of the earth, and the distance of separation between us and the center of mass of the earth is usually pretty constant. The radius of the earth is about 4,000 miles. Since our separation from the center of the earth is very large, slight changes in altitude will make only very small changes in the total distance of separation. Although the earth's gravitational force acting on you on the summit of Mount Everest will be a little less than that acting on you at sea level, the difference will be less than a pound. (To lose weight, increasing altitude is much less effective than dieting or exercise.) For positions on or near the surface of the earth, we can consider the downward force of gravity on a particular mass, like your body or some object you want to levitate, to be *very nearly constant*.

Even if you are orbiting in the space shuttle about 200 miles above the earth's surface, that increases your distance from the center of the earth by only 5% (from 4,000 to 4,200 miles) and decreases the gravitational force on you by only about 10% (inverse square—the square of 1.05 is about 1.10). So the earth is exerting on you and the other astronauts in the shuttle about 90% of the gravitational force that it exerts on you when you're on the earth. It's not "zero gravity" as it sometimes is mistakenly called. The gravitational field of the earth, although it decreases with distance from the earth, extends to infinity. In the shuttle, it's just slightly reduced gravity.

But if that's true, why can you and your fellow astronauts float around so weightlessly? Because you're in *free fall*. You are falling like a skydiver but without the air resistance. And also unlike a skydiver, you're not falling straight down. The forward velocity of the shuttle in orbit keeps it and you falling *around* the earth, not into it. If you want to experience free fall without a ride in the shuttle, some amusement park rides offer 2 or 3 seconds of nearly free fall (thoughtfully followed by rapid deceleration before you hit the ground). Special airplane flights offer, several times during the flight, up to 25 seconds of nearly free fall. (NASA calls their plane Weightless Wonder, but for those with weak stomachs, it has become known as the "vomit comet.") The scenes in the movie *Apollo 13* where Tom Hanks and other astronauts are floating around weightlessly were filmed in such a plane. But in this book we will be considering objects that are very near the surface of the earth and not in free fall. In fact, with maglev, we want to use magnetic forces to keep objects from falling.

Before we consider our main object of interest, the magnetic force, we should say a few words about its close cousin, the *electrostatic force,* another aspect of the electromagnetic force. The electrostatic force can operate at large distances between objects if one or both objects have a net electrical charge. When an object is uncharged or electrically neutral, as most objects usually are, the negative charge of the electrons is balanced by the positive charge of the atomic nuclei. However, it is often easy to create a surplus or deficiency of electrons in an object, say by rubbing a rubber comb through your hair (or rubbing it with wool if you no longer have hair). The rubber comb picks up some extra electrons from your hair or the wool, gets a net *negative* charge, and is now able to attract light bits of paper. (You should try this on a day when the air is fairly dry. If the air is very humid, water in the air tends to drain the comb of its excess charge.) Although the paper is uncharged, the net negative charge on the comb repels electrons in the paper away from the surface and the paper surface facing the comb gets a net positive charge, resulting in attraction to the comb. Another common experiment is to rub a glass rod with a bit of silk. The glass will *lose* a few electrons to the silk and get a net *positive* charge, and it now can also attract light bits of paper. Here the net positive charge of the glass rod attracts electrons and the surface of the paper facing the glass rod gets a net negative charge. Objects with a net charge, either negative like the rubber comb or positive like the glass rod, can attract uncharged objects like the bits of paper simply by causing the motion of some electrons in the paper. Transfers and motions of electrons among clothes in the dryer produce the phenomenon of "static cling" referred to in Chapter 1.

A charged object attracts uncharged objects, but if you have two objects, each with a net electric charge, they can either attract or repel depending on their relative sign. *Like charges repel, unlike charges attract.* If free to move visibly in response to weak forces, say if hung by thin threads, two rubber combs rubbed with wool (each with a net negative charge) will repel each other, as will two glass rods (each with a net positive charge) rubbed with silk, but a rubber comb and a glass rod will be attracted to each other. These electric forces of attraction or repulsion between charged objects act at large distances of separation, and Frenchman Charles-Augustin de Coulomb showed that the decrease in force with increasing separation followed the same mathematical law that Newton postulated for gravitational forces—the inverse-square law. "Coulomb's law" of electrostatic forces states that electric forces between two objects with net electric charge decrease as the *inverse square* of the distance between them.

It's instructive to set Newton against Coulomb and compare the relative strengths of gravitational and electrostatic forces. Two protons, elementary particles of positive charge, exert on each other both an attractive gravitational force and a repulsive electrostatic force. But it is not a close contest. The electrostatic force of repulsion between the two protons, at any distance, is stronger than the gravitational force of attraction by a factor of 10^{36} – 1 followed by 36 zeroes. (That's a trillion trillion trillion!) Nevertheless, despite the relative weakness of gravitational forces compared to electromagnetic forces, they are very noticeable for those of us who live on the surface of a very large mass like the earth. The famous physicist Richard Feynman was once speaking at a physics conference on the relative weakness of gravity. "The gravitational force is weak," he said, "In fact, it's *very* weak." As luck would have it, at that very instant a loudspeaker broke loose from the ceiling and crashed to the floor. Feynman quickly added, "Weak–but not negligible."

Magnetic Forces

Having discussed the basics of the gravitational force, which is essentially constant but definitely "nonnegligible" near the surface of the earth, we now turn to the forces that are used in maglev to combat the pervasive and nonnegligible force of gravity—magnetic forces. The source of gravitational forces is mass. The source of electrostatic charges is electric charge. The source of magnetic forces is also electric charge, but electric charge *in motion,* even though you usually cannot see the motion of the charge, which can be electric current or can be on the atomic scale in the form of spin or orbital motion of electrons. In fact, it is the motion-based connection between electricity and magnetism that led Einstein to his theory of special relativity, which deals with relative motion. But for this book, we will ignore relativity and focus on situations in which the forces we are using to counter gravity can simply be considered as magnetic forces.

Here it will be useful to remind you of several basic things about magnets and magnetic forces you probably learned many years ago, but several of which you are likely to have forgotten if you are not a practicing scientist or engineer. I introduced them in my earlier book, *Driving Force,* as ten "facts about the force." (If you have read and memorized *Driving Force,* you can skip this section.)

Fact 1: If free to rotate, permanent magnets point approximately north–south.

That's what compass needles do and what other magnets will do if they are free to swing in response to the weak roughly north–south magnetic field of the earth. The end of the magnet that points north we call the *north pole* (north-seeking pole) and the end that points south we call the *south pole*. The ability of magnets to identify directions was known in China well over 2,000 years ago, but became known in Europe only about 1,000 years later.

Note that the earth's magnetic field causes the compass needle to rotate, that is, exerts a *torque* on the compass, but does not produce any net force to move the compass needle southward or northward. Because the earth's field is essentially constant over the length of the compass, the magnetic forces exerted by the earth's field on opposite poles of the compass needle are equal and opposite—no net force.

Fact 2: Like poles repel, unlike poles attract.

The north pole of one magnet will repel the north pole of a second magnet, but will attract its south pole. Unlike gravitational forces, *magnetic forces can be either attractive or repulsive.* Many students I have talked with about magnets, from first-graders to engineering majors at MIT, seem to find repulsive magnetic forces more fascinating than attractive forces. And we will find that many techniques of magnetic levitation are based on repulsive forces.

Fact 3: Permanent magnets attract some things (like iron and steel) but not others (like aluminum, copper, wood, or glass).

This selectivity is one of the most fascinating, and most useful, aspects of magnetic forces. It is used a lot in industry, and you can use it yourself to separate out the tiny iron-rich particles that are present in sands and soils and even some fortified breakfast cereals.

Fact 4: Magnetic forces act at a distance and can act through non-magnetic barriers (if not too thick).

This is also very useful, not only to engineers but also to magicians who can achieve mysterious effects with hidden magnets. It also adds to the aesthetic effect of some maglev devices.

Fact 5: Things attracted to a permanent magnet become temporary magnets themselves.

When touched by a permanent magnet, a steel paper clip becomes magnetic enough to attract a second paper clip; it becomes a "temporary magnet." If your magnet is strong enough, you can hang a long chain of paper clips under it. These first five "facts about the force" were known even in ancient times. In one of Plato's early dialogues, he has Socrates describing Fact 5 in detail, although he discussed a chain of iron rings, not steel paper clips. (Socrates, via Plato, was not just conveying scientific facts. He was using this familiar property of magnets as a metaphor for the ability of poets to inspire others.)

The next three facts were not known until the nineteenth century, when the close connection between electricity and magnetism was discovered. These may be less familiar to you than the first five, but some will be very important to the topic of magnetic levitation.

Fact 6: A coil of wire with an electric current flowing through it becomes a magnet.

It was in 1820 that Danish physicist Hans Christian Oersted discovered that an electric current flowing down a wire moved a nearby compass needle. In a wire wound into a coil (e.g., shaped like a bedspring or a Slinky), the magnetic fields from the various turns of wire add together, and the coil becomes an *electromagnet* with a north pole on one end of the coil and a south pole on the other (Figure 2). Increasing the current increases the strength of the electromagnet, and if the current is reversed, the poles reverse. Permanent magnets are wonderful things, but electromagnets have the distinct advantage that their magnetic fields can be changed by changing the current. Permanence is admirable, but flexibility also has its advantages. Magnetic fields of electromagnets can be turned on and off, and alternating magnetic fields can be produced by alternating currents. That's often very useful, and is especially so in many levitation systems.

Fact 7: Putting iron inside a current-carrying coil greatly increases the strength of the electromagnet.

Iron can be magnetized (become a temporary magnet) by exposure to a permanent magnet (Fact 5) or an electromagnet (Fact 7).

And an iron core inside an electromagnet can increase the strength of its magnetic field by factors of thousands. Many electromagnets in engineering use, including many of those in devices discussed in this book, are iron-core electromagnets. Figure 2 shows, schematically,

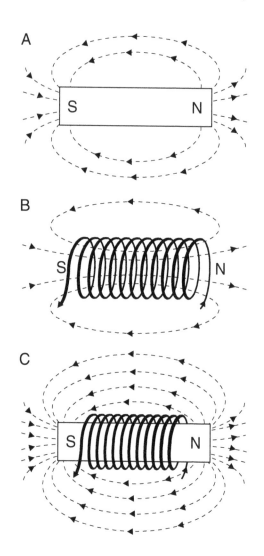

Figure 2. Magnetic field patterns around (top) a permanent magnet, (middle) an air-core electromagnet, that is, a current-carrying coil of wire, and (bottom) an iron-core electromagnet; the iron amplifies the magnetic field produced by the coil.

magnetic fields produced by a permanent magnet, an air-core electromagnet, and an iron-core electromagnet. We will encounter all three magnet types in later chapters.

Fact 8: *Changing* magnetic fields induce electric currents in copper and other conductors.

Shortly after Oersted discovered that you can use electric currents to produce magnetism, Michael Faraday and Joseph Henry independently discovered that you can do the reverse—use magnets to produce electric currents, the phenomenon we call *electromagnetic induction.* Faraday found he could induce currents simply by moving a magnet in the vicinity of a conductor. That's still how we generate most of our electricity today. Fact 8 may be unfamiliar to you, but we'll later devote one whole chapter to it (Chapter 5), since using electromagnetic induction is one of the important ways to produce magnetic levitation. Among other things, it's what allows some Japanese maglev trains to lift off the tracks and break world speed records.

The final two of the ten "facts about the force" deal directly with magnetic forces, and since in magnetic levitation we'll be using magnetic forces to counter gravitational forces, we should include them, even though they may be totally new to you. We'll refer to Fact 10 a few times in later chapters and to Fact 9 at least once. And they both are pretty basic to the essence of magnetism and its uses. They may be totally unfamiliar to you, but when we use them in this book, we'll remind you of what they are.

Fact 9: A charged particle experiences no magnetic force when moving parallel to a magnetic field, but when it is moving perpendicular to the field it experiences a force perpendicular to both the field and the direction of motion.

In freshman physics courses, Fact 9 is essentially introduced as the *definition* of a magnetic field. Electric fields are defined in terms of the force they exert on even stationary charged particles, and magnetic fields are defined in terms of the force they exert on *moving* charged particles. We won't use Fact 9 much in this book, but the fact that magnetic fields produce forces perpendicular to the direction of motion of charged particles produces circular or curved paths of the particles in many large and small devices, including the electrons moving in circular paths in the magnetron that generates microwaves in your microwave oven. And it's how the earth's magnetic field directs charged particles emitted from the sun to the earth's poles, thereby both protecting us from some radiation and giving us the northern (and southern) lights.

Fact 10: A current-carrying wire in a perpendicular magnetic field experiences a force in a direction perpendicular to both the wire and field.

Fact 10 is a close cousin of Fact 9 and deals with charged particles (electrons) moving through a conducting wire (i.e., electric current) rather than through free space. In this book, when we are dealing with moving electric charges, they will mostly be moving not through free space but through conducting metals, as in the wires of a current-carrying electromagnet. So Fact 10 will be more important to us than Fact 9. We'll remind you of it when we need it.

In *Driving Force,* I used Fact 10 to explain how magnetic forces on current-carrying wires in *speakers* convert alternating electric currents into alternating forces, producing alternating motion of a diaphragm that generates alternating compression and expansion of air (i.e., generates sound). Fact 10 can also be used to explain how magnetic forces acting on current-carrying wires convert currents into motion in *motors,* the other most common use of magnets. But in *Driving Force,* I instead discussed a motor in terms of Fact 2, the force between magnetic poles. Sometimes you can think of an electromagnet as a magnet with north and south poles, but sometimes it's more helpful to think of it as current-carrying wires. It's both.

There are lots of perpendiculars in Fact 9 and its close cousin Fact 10, and for full application, they involve some three-dimensional thinking and one of the venerable "right-hand rules" popular (at least to the professor) in physics courses. Those rules often produce the amusing sight of many physics students, during exams, pointing the thumbs and index and middle fingers of their right hands in mutually perpendicular directions to figure out the direction of magnetic forces. We'll spare you that challenge.

Phew! I've already thrown a lot of magnet science at you, and we're only in Chapter 2. Take a deep breath. You probably knew the first five facts before picking up this book, even though you may not have thought about them much lately. So a reminder may have been helpful, and we'll remind you again if and when we need to. Facts 6 and 7 you probably learned in school through the traditional experiment of wrapping a wire around a nail, attaching the ends of the wire to a battery, and finding that the nail was now a magnet and could pick up paper clips. But that was a long time ago, so I suspect that reminder was even more helpful. And Facts 8, 9, and 10 may be total strangers to some readers. Fear not. In the coming chapters, we will not be assuming that you already know all these things. As noted earlier, we will be devoting all of Chapter 5 to discussing Fact 8 (electromagnetic induction). And when we need Facts 9 or 10 (mostly 10) in later chapters, I'll be sure to remind you of them. When you see a fact about magnets and magnetic forces applied to a real physical situation, it will be a lot more meaningful than considering it in the abstract.

Magnetic Materials

I teach in the materials science department at MIT, so it's appropriate that before discussing the fundamentals of magnetic levitation, I should also say a few words about magnetic materials, and in particular, about the one element that is at the center of the world of magnetism—iron.

Iron would be a very special element even if it were not magnetic. Of all the hundred-odd elements, iron has the most stable nucleus, which results in iron being the most common metal in the universe. And, by mass, it is also the most common of all the elements in our planet earth. Oxygen, aluminum, and silicon are a bit more common than iron in the earth's crust; because iron is heavier, much of it is in the earth's core. But iron is much easier to separate from its oxides than silicon and aluminum. And in the form of steel, it is mechanically strong, formable, and inexpensive, making it of extreme importance as a structural material.

And iron's magnetic! Its immediate neighbors in the periodic table, cobalt and nickel, are also magnetic, but they are much more expensive than iron, so they are much less used. As we discussed in Facts 5 and 7, iron becomes magnetized in the presence of perma-nent magnets and as cores in electromagnets, and the use of iron as a "temporary magnet" is important in many applications.

It also turns out that iron is a major chemical component of most permanent magnets. Lodestones, the only magnets known to Plato and his contemporaries, are made of the mineral magnetite, an iron oxide found naturally around the world. Once scientists and engi-neers learned to make steels (iron containing carbon and other ele-ments), steel magnets surpassed lodestones, and they played a central role in the development of the electrical industry in the nineteenth century. In the 1930s, alnico magnets (alloys of iron with aluminum, nickel, and cobalt) surpassed the steels, and they are still sometimes used today. Alnico magnets helped the Allies defeat Hitler through magnetrons and microwave radar, but today most permanent mag-nets used are either the "ceramic" ferrite magnets (iron oxides contain-ing barium or strontium) or the "rare earth" neodymium magnets (an iron compound with neodymium and boron).

Thus the overwhelming majority of permanent magnets used in the past and those used today contain lots of iron. The ferrites are much cheaper than neodymium magnets and are therefore more used; for example, most refrigerator magnets contain fine particles of ferrite magnets immersed in plastic or rubber. But neodymium mag-nets are much more powerful than the ferrites and are used wherever superior magnetic properties are more important than cost, particu-larly in applications where size and weight are critical, as in laptop computers and earphones. And they are used in many levitation devices because they can produce greater magnetic forces than ceramic magnets.

Neodymium, an important component of today's most powerful permanent magnets, is element number sixty, one of the so-called "rare earth" elements that usually appear near the bottom of charts of the periodic table of elements. Neodymium itself actually is not very rare and is more plentiful in the earth's surface than more familiar elements such as lead and tin. However, it is relatively difficult to separate from its ores, making it fairly expensive. And over 90% of the earth's known supply of rare earths is in China; half comes from a single mine in Inner Mongolia. In 1992, Chinese president Deng Xioping bragged, "The Middle East has oil, China has rare earths." There are other important industrial uses for rare-earth elements, but the importance of neodymium in particular to today's high-strength magnets makes some analysts worry that China is developing a stranglehold on the world market for magnets, and neodymium magnets are important in much of modern technology. They are especially important in computer hard drives and also are key to several growing markets for environmentally friendly "green" technology, including generators for wind turbines and electric motors for hybrid and electric cars. One of the biggest users of neodymium today is the Toyota Prius. In 2009, the Chinese Ministry of Industry and Information Technology announced tighter restrictions on exports of several rare-earth metals, including neodymium, and the announcement instantly produced greatly increased interest in the mining of rare-earth ores in the Western world (and a strong increase in the stock prices of mining companies in the business). Some darkly accused China of wielding the rare earths as a "twenty-first century economic weapon." Neodymium and the other rare earths are not yet as important to the United States as Middle Eastern oil, but if we ever carry through on our long-stated goal of reducing our dependence on foreign oil, the balance will shift.

We'll find that neodymium permanent magnets and iron-core and air-core (iron-free) electromagnets all play important roles in the world of magnetic levitation. And most of the above "facts about the force" will be important to our discussions about maglev. Now that we have reviewed the basics of gravitational and magnetic forces and magnetic materials, it is time to consider the basics of using magnetic forces to combat gravitational forces, that is, to consider the fundamentals of magnetic levitation.

Maglev — A Balance of Forces

Poles Apart

Most dictionaries define "magnet" as a "body that attracts iron." As noted in Fact 3 in the previous chapter, although magnets attract iron, they do not attract copper or aluminum, or most other things—plastics, glass, wood, and so on. My granddaughter discovered this shortly after she learned to walk by removing a magnet from our refrigerator door and trying to stick it, usually unsuccessfully, to other surfaces around our house. Magnetic forces are very selective, a property that is not only fascinating but also very useful in separating magnetic materials from nonmagnetic materials. For example, since hay is nonmagnetic and most needles are made of steel, with the use of a magnet it is really very easy to "find a needle in a haystack."

The forces between two magnets are even more interesting, since they are sometimes attractive and sometimes repulsive. Each magnet has at least two poles, a north pole and a south pole, defined by which way they point in the earth's magnetic field if free to rotate, as in a compass (Fact 1). As I demonstrated to the wonder of my granddaughter's first-grade class recently, south and north poles attract each other, but two south poles, or two north poles, repel each other. *Unlike poles attract, like poles repel* (Fact 2).

Disc and ring magnets are usually magnetized in a direction through their thickness, so that one face is a north pole and the opposite face a south pole. With two ring magnets on a pencil, the repulsive force between two like poles facing each other, say north–north, can lift the upper magnet against the downward force of gravity and produce the simplest form of magnetic levitation (Figure 3). The downward gravitational force on the upper magnet is balanced by the upward magnetic force of repulsion between the two magnets. (The two magnets in Figure 3 are ferrite magnets. If you instead use two neodymium ring magnets, the repulsive forces are stronger and the levitation height is considerably higher.)

Here the repulsive force between the two magnets is directed perpendicular to the plane of the magnets, pushing the two north poles apart. The direction of magnetization of the magnets and the direction of the repulsive force between the magnets are both vertical. If instead you put two ring magnets on a table with the two north poles each facing upward, the magnets also repel each other in the *sideways* direction—the direction parallel to the plane of the magnets and perpendicular to the direction of magnetization.

This sideways repulsion is the secret of the maglev toy called Revolution (Figure 4), in which magnets in the base, magnetized in the horizontal direction, repel two similarly magnetized disc magnets above them in the rotating part, producing levitation. In both Figure 3 and Figure 4, the magnetic repulsive force producing levitation is directed vertically upward against gravity, but in Figure 3 it is parallel to the direction of magnetization, while in Figure 4 it is perpendicular to the magnetization, parallel to the surface of the two disc magnets. The seeming defiance of gravity is more visually striking in the latter case, since the magnets in Revolution are less constrained and closer to full contact-free levitation than the magnets on the pencil.

However, in neither of these cases are the levitated magnets totally free of contact. Touch was not involved in the upward push of magnetic repulsion that opposed the downward pull of gravity, but was necessary to keep the position of the levitated magnets stable. If you remove the pencil in Figure 3 and try to float the upper magnet above the lower one, it will move sideways or flip over until opposite poles mate. If you remove the glass plate in contact with one end of the rotator in Figure 4, the rotator will lurch forward and fall. As

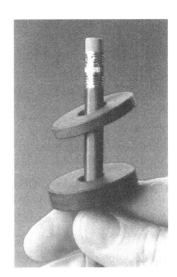

Figure 3. When two ring magnets are placed on a vertical support with like poles facing, the upper magnet is levitated by the repulsive force between the two magnets.

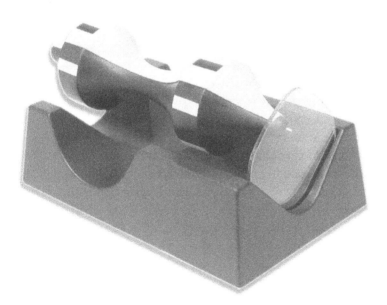

Figure 4. The horizontal rotating part (rotor) of the "Revolution" maglev toy is held up by repulsive forces between like poles of ring magnets in the rotor and pairs of triangular magnets beneath each ring magnet. The rotor can turn for several minutes because the only friction is between the steel-pointed end of the rotor and the glass plate that blocks its horizontal motion.

many children and even adults have found with their futile attempts over the years, you can't simply use the repulsive force between like poles of two permanent magnets to levitate the upper magnet freely in space without it being in contact with something, like the pencil in Figure 3 or the glass plate in Figure 4. Way back in 1842, Samuel Earnshaw, an English clergyman and mathematician, proved mathematically that stable contact-free levitation by forces between ordinary stationary magnets alone is impossible. So you can stop trying. But as we will see in the coming chapters, there are several ways to get around the Reverend Earnshaw's proof and achieve contact-free magnetic levitation. And for many applications, a little contact doesn't hurt. In the Revolution, the friction between the glass plate and the sharp metal tip of the rotating part (rotor) is pretty small, and once started, the rotor can spin for several minutes before slowing to a gradual stop.

Force at a Distance

Another thing about magnetic forces that children learn pretty quickly when playing with magnets is that the forces between magnets fall off rapidly with distance between them. Gravitational and electrostatic forces also decrease with distance, and according to Newton and Coulomb, they decrease as the *inverse square* of distance of separation. But magnetic forces *decrease more rapidly* than as inverse square of distance—because magnets always have at least two poles.

You may recall an experiment they often do in schools. Take a long bar magnet, with a north pole on one end and a south pole on the other, and break it in two. You find that now each of the two resulting magnets has both a north pole and a south pole. Two new poles have appeared on the new surface created by the break. All magnets have at least two poles, a north pole and a south pole— magnets are *dipoles,* not monopoles. If magnetic monopoles existed, they would attract or repel each other with an inverse-square law.

But for ordinary magnets, it is safe to assume that they have at least two poles. So the forces between magnets (i.e., between magnetic dipoles) are more complex than the inverse-square laws of gravitational and electrostatic forces.

One important result of the fact that magnets are dipoles is that if a magnet is in a uniform magnetic field, the forces on opposite poles are equal and opposite, and the magnet feels no net force, as noted in the previous chapter. All it experiences is a *torque* that attempts to rotate it and align it, like a compass needle, with the local magnetic field. But there *can* be a net force on a magnet if the field is not uniform—if it changes with distance. A piece of iron is attracted to a magnet because the field is higher near the magnet than away from it. Where the magnetic force is repulsive, as in Figure 3, the upper magnet is pushed upward because the field from the lower magnet is lower there. We call that change of the magnetic field strength with distance the *gradient* of the field, in analogy to the gradient or slope of a hill, which is the change of height with horizontal distance. (For those who remember their calculus, the field gradient is the *derivative* of the field with respect to distance, but I promise not to mention calculus again.) If the field changes with distance, as it will in the vicinity of another magnet, the force on one pole will be greater than the force on the other pole, and there will be a net force.

Now consider those two ring magnets on the pencil in Figure 3. The north pole of the top magnet is repelling the north pole on the bottom magnet. But each of the magnets has another pole! The south pole on the top magnet is attracting the north pole of the bottom magnet, and the south pole of the bottom magnet is attracting the north pole of the upper magnet. This decreases the net repulsive force between the two magnets, and the result is that, at large separations, the repulsive force between the two magnets varies not as the inverse square, but as the *inverse fourth power* of the distance between them. If you double the distance of separation, the force decreases by a factor of sixteen—two to the fourth power ($2 \times 2 \times 2 \times 2$).

That's quick! The force–distance relation all the way from small separations to large separations is more complex—I won't burden you with it. The major point to remember is that the *magnetic forces between magnets decrease very rapidly with distance of separation.*

The force of attraction between a magnet and a piece of iron decreases much faster yet because it depends not only on the gradient of the magnetic field, but also on the strength of the local field of the magnet acting on the iron, which determines the degree of magnetization of the iron. The net effect is that at large separations, the force between a permanent magnet and a piece of iron (always attractive) decreases as the *inverse seventh power* of separation! With the inverse-square laws of gravity and electrostatics, doubling the separation decreases the force by a factor of four (2^2). With the inverse-fourth-power law for the force between two distant magnets, doubling the separation decreases the force by a factor of sixteen (2^4). With the inverse-seventh-power law for the attractive force between a magnet and a piece of iron, doubling the separation decreases the force by the remarkable factor of 2^7, or 128!

The Bottom Line: Since the force of gravity remains nearly constant near the surface of the earth, but magnetic forces decrease very rapidly with distance of separation, magnetic levitation gaps will in general be very limited. To a degree, the "reach" of a magnet's field can be increased by increasing the magnet's size, but to achieve levitation gaps greater than a few inches, you'll need magnets that are both very strong and extremely large (and therefore extremely expensive). Sorry about that.

The Floating Island of Laputa

For the reasons outlined above, magnetic levitation heights are generally limited to distances of inches or less rather than miles. But in fiction like Jonathan Swift's satirical novel, *Gulliver's Travels* (1726), imagination is not limited by facts. In Swift's novel, after Gulliver had visited the tiny people of Lilliput and the giants of Brobdingnab, he next visited Laputa, a magnetically levitated "Island in the Air" (Figure 5). Swift described this remarkable but imaginary island in considerable detail. It was circular with an area of 10,000 acres, about 70% of the area of the island of Manhattan, and achieved levitation to heights of up to 4 miles by means of a giant lodestone, a natural magnet. The magnet was shaped somewhat like a cigar, about 6 yards long and about 3 yards in diameter. As described by Swift, the magnet was

> endued at one of its Sides with an attractive Power, and at the other with a repulsive. Upon placing the Magnet erect with it attracting End towards the Earth, the Island descends; but when the repelling extremity points downwards, the Island mounts directly upwards . . . When the Stone is put parallel to the Plane of the Horizon, the Island standeth still; for in that Case, the Extremities of it being at equal Distance from the Earth, act with equal Force, the one in drawing downwards, the other in pushing upwards; and consequently no Motion can ensue.

Swift was clearly aware of one basic fact about magnets—like poles repel, unlike poles attract (Fact 2). Although he did not describe the forces levitating Laputa in terms of north and south poles, it can be assumed that the lodestone was magnetized along its length, with a north pole on one end and a south pole on the other. Apparently Balnibarbi, the land above which Laputa levitated, contained underneath it a gigantic permanent magnet, a "Mineral which acts upon the Stone in the Bowels of the Earth." And Balnibarbi's gigantic underground magnet was predominantly of one pole facing upward, so that it repelled one end of Laputa's lodestone and attracted the other. But among the various unrealistic properties of Laputa was its ability to stably maintain altitude when the lodestone was horizontal. In such an orientation, there would be no net upward levitation force to oppose gravity. It would not hover at a fixed altitude; it would fall. And the vertical magnetic field from Balnibarbi would

Figure 5. Gulliver wondering at the magnetically levitated island of Laputa.

exert a huge torque on the horizontal lodestone, tending to tilt the lodestone. But according to Swift, tilting of the lodestone produced motion of Laputa at an angle to the vertical, motion with a horizontal component that allowed Laputa to travel to other parts of Balnibarbi. That also wouldn't work.

Science-fiction and popular-science author Isaac Asimov in 1980 published an annotated version of *Gulliver's Travels* in which he analyzed Swift's description of magnetic forces in detail. He concluded, "In short, Swift's mechanism to keep Laputa aloft and in motion wouldn't really work, but I suppose that's of no surprise to anyone. The explanation sounds scientific and impressive and that's all that counts." Swift's motive in describing the impossible motions of Laputa was simply to satirize scientists and their writings, not to provide a scientifically accurate explanation of magnetic levitation. In addition to the inconsistencies noted above, to levitate an island of the weight of Laputa through forces on a lodestone of that size would require an impossibly strong repelling magnet, many miles wide, in the ground below. And it would require some means to counter Earnshaw's rule and produce stable magnetic levitation. But *Gulliver's Travels* is still a fascinating read, and among its pleasures is the remarkable but impossible floating island of Laputa.

Unlike Swift, most fiction writers featuring levitation, like J. K. Rowling of the Harry Potter books, do not specify that magnetic forces are the source of the levitation. However, the writers of the *Dick Tracy* and *Spiderman* comics of the 1960s explicitly did. Dick Tracy and his colleagues flew around town in one-man maglev devices that, like Laputa, could achieve altitudes of miles. Chester Gould, the creator of Dick Tracy, was so impressed with the potential of magnetic forces that his comic strip frequently stated, "The nation that controls magnetism will control the universe." And the "magnetic antigravity device" used by Spiderman's adversary Vulture also reached impressive heights. Spiderman developed an "antimagnetic inverter" to weaken Vulture's mysterious maglev device, but Vulture returned in a later issue with an improved version. Beyond asserting that their characters achieved levitation through magnetism, the authors of *Dick Tracy* and *Spiderman* provided no scientific details that could be critiqued by Isaac Asimov. And they, like Jonathan Swift, conveniently ignored the decrease of magnetic forces with distance of separation and thereby achieved substantial heights of levitation. Writers of nonfiction like myself are more constrained.

Stability and Degrees of Freedom

The upper ring magnet in Figure 3 floats above the lower ring magnet at a fixed and stable distance of separation. The downward gravitational force on the upper magnet is essentially constant, but the upward repulsive magnetic force decreases rapidly with distance of separation, and the magnet rests at a separation where the upward and downward forces are equal and opposite. There's a balance of vertical forces. No net vertical force, no vertical motion. Now if you displace the upper magnet a little bit upward from that position, the upward repulsive magnetic force will decrease and become less than the downward gravitational force, which is unchanged. The net force will be downward, pulling the magnet back to its original separation. If instead you were to displace the upper magnet downward, the upward magnetic force will increase to more than the downward gravitational force. The net force will now be upward, pushing the magnet back to its original separation. Thus displacement in either direction results in a *"restoring force"* acting to return the magnet to its original position where the upward and downward forces are equal and opposite. We say that there the upper magnet is in a position of *stable equilibrium* (with respect to vertical displacement). For small displacements from a position of stable equilibrium, the restoring force increases linearly with displacement, and in analogy to the forces resulting from stretching or compressing a spring, the ratio of restoring force to displacement is called the *spring constant*.

Suppose we instead reversed one of the magnets on the pencil so that they *attract* each other. Imagine then holding the upper magnet fixed and allowing the lower magnet to move up and down along the pencil. It would be possible to find a separation between the two magnets where the downward gravitational force and an upward magnetic force, now attractive, are equal and opposite. As in the repulsive case, there's a balance of vertical forces in this position and no net force. But if you now were to displace the lower magnet a little bit upward toward the upper magnet, the upward attractive magnetic force would increase, the net force would be upward, and the magnets would snap together. If instead you were to displace the lower magnet downward, the upward magnetic attractive force would decrease, the net force would be downward, and the lower magnet would fall. Unlike the repulsive case, displacements from the position of zero net force lead not to restoring forces, but to net forces that move the magnet *away* from that position. That position is *unstable.* Instead of net restoring forces, there are net *destabilizing forces.* Since gravitational forces are constant but magnetic forces decrease rapidly with distance of separation, it is often easier to achieve stable levitation with repulsive forces than with attractive forces. If you're attracted to something, it's hard to keep your distance!

But even in repulsion, we surely needed that pencil in Figure 3. By confining the magnets to the pencil, we could reach stable equilibrium of vertical forces. But if we tried to levitate one ring magnet above the other without a pencil, the upper magnet would spontaneously move sideways or tip over. It was stable, with displacements resulting in restoring forces, only for displacements in the vertical direction. But it was unstable with regard to sideways displacements or tipping displacements. If the lower ring magnet were made bigger than the upper one, the changed magnetic field distribution could stabilize the upper magnet against sideways displacements, but it would still be unstable to tipping motions. Full stability is not possible. Samuel Earnshaw proved mathematically, and many have proved by experiment, that you can't levitate one stationary magnet above another without some contact. Earnshaw's rule is an important limitation to the possibilities of magnetic levitation.

Thinking big, a Dutch company called Universe Architecture has recently designed a full-size "floating bed" based on repulsion between like poles of large arrays of permanent magnets, presumably neodymium. But to satisfy Reverend Earnshaw, the floating bed is mechanically stabilized by angled cables running from each of the four corners of the floating bed to the floor. Here the cables play the role of the pencil in Figure 3 and prevent the bed from flipping over, which would be disconcerting to a sleeper if it happened in the middle of the night. The magnetic floating bed is an intriguing use of Fact 2, but the bed weighs several tons and is priced at over a million euros. There have been no customers yet.

Levitation of the two disc magnets in the rotor in Figure 4 by magnets in the base is more complex and more interesting. As with the magnets on the pencil, upward repulsive magnetic forces balance downward gravitational forces to achieve stability in the vertical direction. Displace the rotating part (the rotor) a bit upward, and it will return to where it was. Push it a bit downward, and it will return to where it was. There are vertical restoring forces. But it's much more stable than that. If you move the rotor a bit sideways, it will also return to where it was. In the base below each disc magnet of the rotor are two triangular magnets, one on the left side and one on the right side, each magnetized in the same direction as the ring magnets, that is, horizontally along the rotor axis (Figure 6). They provide a *"magnetic well"* that exerts repulsive forces on the disc magnet (represented by arrows in Figure 6) roughly perpendicular to their hypotenuse (the long side of the triangle), thus with both vertical and horizontal components. The vertical components add together to provide a net vertical repulsive force upward on the disc magnet. The horizontal components, in contrast, are in opposite directions and cancel each other when the disc magnet centers between them. However, when the disc is displaced to one side, it becomes closer to the triangular magnet on that side. That repulsive force therefore increases while the repulsive force from the other triangle decreases, and the net horizontal force provides a *restoring force* against sideways displacements.

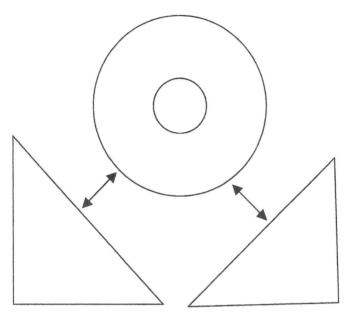

Figure 6. Schematic of the magnet positions in the Revolution toy in Figure 4. Each of the two ring magnets on the rotor is supported by a pair of triangular magnets on the base. Direction of magnetization is normal to the figure. Repulsive forces between the magnets (indicated by arrows) both lift the ring magnet and resist lateral displacements.

Now try some tipping displacements. With one end of the rotor fixed in position on the glass plate, displace the other end up or down a bit so that the rotor is no longer horizontal, and the rotor will return to horizontal. Displace that end sideways so that the rotor is no longer perpendicular to the glass plate, and it will return to perpendicular. The repulsive forces between each of the disc magnets in the rotor and the two triangular magnets in the base below each of them provide *restoring forces* and *stable equilibrium* not only for vertical displacements, but also for sideways displacements and for tipping displacements in two different directions.

The Revolution device in Figure 4 (patent filed in 1991 by Gary Ritts of California) is a far more sophisticated maglev device than the two ring magnets on a pencil in Figure 3. But it still falls short of our goal of complete contact-free levitation. Although stable against displacements in two directions of motion (vertical and sideways) and two axes of tipping, and allowing free rotation about the axis of the rotor (neutral stability around that axis—no restoring or destabilizing forces), the rotor would not be stable in the horizontal direction along its axis without the glass plate. The disc magnets are placed slightly closer to the glass plate than the base magnets, producing a small magnetic force pressing the rotor against the glass plate and keeping the rotor from moving backward.

It is time to introduce an important term often used in discussing maglev devices (as well as in other areas of science and technology): *degrees of freedom*. If we are to levitate an object stably in space, it should be stable against displacements of *six* different sorts. It should be stable against displacements in each of three mutually perpendicular directions, say north–south, east–west, and up–down (or along x, y, and z axes for those more mathematically inclined). And it should also be stable against rotations about each of those three perpendicular directions. In some applications, it will also be desirable to have all those six degrees of freedom—three position variables and three rotation variables—not only stable, with restoring forces resisting displacements, but also fully controllable with magnetic forces.

With the two magnets on a vertical pencil in Figure 3, the pencil has removed any possibility of displacements in either of two perpendicular horizontal directions and substantial rotations about either of these horizontal directions. (As you can see in the figure, the upper magnet is trying to tip over, but can't.) The pencil has removed four degrees of freedom for the upper magnet and limited it to only two—vertical position along the pencil and rotation about the pencil. If the magnets are fully circular in symmetry, rotation about the pencil has no effect on magnetic forces (neutral stability). Only one of the six degrees of freedom, vertical position along the pencil, is stably controlled by magnetic forces.

In contrast, the glass plate in the Revolution device of Figure 4 has constrained only one degree of freedom of the rotor—displacement along its axis. As we have seen, magnetic repulsive forces between the two ring magnets and the pairs of triangular base magnets below each of them keep the rotor stable against both vertical and sideways displacements and also against tipping rotations about either of those directions. So the rotor is stably held with respect to those four degrees of freedom and has neutral stability against a fifth—rotation of the rotor about its axis. It's unstable to only one of the six degrees of freedom—displacement along the rotor axis. Thus in the Revolution we have come a long way toward full contact-free levitation, but to achieve that, we will have to liberate all six degrees of freedom from constraint by physical contact. We will need some tricks to get around the dictates of the Reverend Earnshaw.

Hint: Earnshaw considered only static magnets and static fields.

Spinning the Levitron

Only a Toy

I have long been a fan of science toys, which are relatively inexpensive but capable of illustrating physical principles in an engaging way. Using them in seminar or lecture can make scientific concepts more accessible and understandable to students from kindergarten to college. The shelves in my MIT office hold numerous science toys, and I have often used the toys related to magnets in a freshman seminar on magnets that I taught there for many years. For example, I often introduced magnetic levitation to the freshmen with the Revolution toy of Figure 4 and then asked them to figure out where the magnets are and in which direction they are magnetized. The resulting discussions and analyses can be quite instructive for the students (and sometimes for me).

Thus I was very excited one day in 1995, shortly after my earlier book on magnets went to press, when I first saw a Levitron, "the amazing antigravity top," in a science-based store. I assumed that the upward levitating force on the top was the same repulsive force between like poles employed in Figure 3, but now there was no pencil needed to constrain four degrees of freedom of the upper magnet. I suddenly realized that in the Levitron, the levitated magnet was a spinning top, and I concluded that the gyroscopic action resisted tipping of the magnet and allowed the top to evade the rule of Samuel Earnshaw (which applies only to static magnets) and achieve contact-free levitation. My immediate reaction was fascination and wonder, coupled with some annoyance with myself: "Darn! Why didn't I think of that?"

But I bought the Levitron and immediately went home to try my hand at achieving the magic of full contact-free magnetic levitation. I soon found it was not that easy to achieve. Within the box were the three main parts of the Levitron: the base, which clearly held a very large magnet; the small top, which contained a disc magnet about an inch in diameter; and a nonmagnetic plastic "lifting plate." There also were a number of tiny brass and plastic washers, a couple of tiny rubber o-rings (to hold the washers on the top), and two thin wooden wedges or shims. The instructions indicated that I should place the lifting plate on the base, spin the top on the lifting plate, and then lift the plate an inch or so above the base until the spinning top floated free. Easier said than done.

The first challenge was getting the top to spin. There were strong magnetic forces between the base magnet and the magnet in the top that mightily kept fighting me. And those forces kept winning. I had to try many times before I could get the top to spin. To make the top easier to spin, the marketers of Levitrons now offer an electrical "Starter" for those "not so nimble-fingered," a category into which I apparently fell.

Once I became intermittently successful in spinning the top, I faced the second challenge—to adjust the weight of the top to the proper value, which was the purpose of all those brass and plastic washers. If when I lifted the plate by an inch or so the spinning top jumped up quickly, it was too light and another washer should be added. If it failed to lift off the plate, it was too heavy and a washer should be removed or replaced with a lighter one. There were washers with weights of 3 grams (about 0.1 ounce), 1 gram, 0.4 gram, 0.2 gram, and 0.1 gram, and to float well, the total weight of the top had to be adjusted, by trial and error, to reach the "correct" weight within about 0.1 gram (1/300 of an ounce). And the instructions warned that the correct weight could change as a result of small changes in temperature. The weight that worked now may not work a few minutes later.

When I had the weight about right, I faced the third challenge, leveling the base. If the base was not perfectly horizontal, the top would fly off to one side or the other. This required use of the thin wooden shims to level the base. I had assumed the floors of my house were close to level, but apparently not. After about an hour fighting all three challenges, I gave up in despair. It was over a week later before I finally worked up the courage to try again, and this time, after about a half-hour of struggle, I had the great satisfaction of seeing the spinning top bobbing around in the air about an inch above the base. Success! It spun there in midair for about a minute, but gradually slowed down, wobbled more and more, and finally flipped over and fell.

I returned to the task, soon got it spinning in midair again (practice helps), and called my wife to show her the wonder of magnetic levitation. She smiled politely, congratulated me on my achievement, and, shortly before the Levitron fell again, asked what it was good for. My fourth challenge became to explain to her the importance of a small spinning top able to fly about an inch above the ground for a minute or so. After expanding a bit about the general wonders of antigravity, I had to admit that, after all, the Levitron was "only a toy."

But it's a toy that the American Association of Physics Teachers reportedly has called "the best science toy in a generation." It's a toy that has sold well for many years now and has led to much-improved versions, various forms of "Super-Levitrons" and Levitron accessories. It's a toy that has generated intense controversy over its invention and development, and a toy that has led several eminent scientists to publish scientific papers explaining its operation. The Levitron may be only a toy, but it's a pretty special and pretty cool toy.

Toy Story

The official birth of what has come to be called "spin stabilized magnetic levitation" can be dated to May 1983 and the granting to inventor Roy M. Harrigan of Vermont of U.S. Patent 4,382,245 for a "levitation device." He reportedly had first succeeded in levitating a spinning magnet in the mid-1970s, but it took several years for skeptical patent examiners to finally grant Harrigan his patent. Perhaps they had been reading Earnshaw. In his patent, Harrigan describes a first or base magnet as a "dish-shaped magnet having a concave surface uppermost" and a second or upper magnet that "may be rotated either manually or by associated apparatus to provide gyroscopic stability" (Figure 7) The upper magnet, that is, the top on top, has "a polar orientation to repel the concave surface of the first magnet." Although the many claims in the patent include another possible form of the base magnet, diagram 1 shows a dish-shaped base magnet with magnetization pointing perpendicular to the curved upper concave surface, with projected extensions of the emerging magnetic fields pointing inward and converging above the base. In the diagram, the upper surface of the base magnet is a north pole, indicating that the lower surface of the magnet in the spinning top would also be a north pole. Like poles repel.

Harrigan probably chose a dish-shaped base magnet to provide a "magnetic well," similar to that provided in one direction by the two triangular base magnets in the Revolution (Figure 6), to provide the top stability against sideways displacements. He presumably reasoned that with a dish-shaped base magnet, sideways displacements of the top from the central axis of the Levitron would bring the top magnet closer to the base, an increased repulsive force, and a restoring force pushing the top back to the center. He also clearly felt that it was important for the magnetization of the base magnet to form "lines of magnetization forming a cone" (converging above the base near the levitation height), and offered another form of base magnet to achieve that: "a plurality of discrete cylindrical magnets having the axis of each disposed at the same angle to a central axis," that is, pointing inward to form the cone-shaped magnetic field pattern he considered necessary. Later mathematical analysis by several scientists revealed that the field conditions for sideways stability for the top magnet are more sophisticated, and later experimental results

United States Patent [19]

Harrigan

[11] **4,382,245**

[45] **May 3, 1983**

[54] **LEVITATION DEVICE**

[76] Inventor: **Roy M. Harrigan**, Bromley Mountain Rd., Manchester, Vt. 05254

[21] Appl. No.: **105,239**

[22] Filed: **Dec. 19, 1979**

Related U.S. Application Data

[63] Continuation of Ser. No. 899,733, Apr. 25, 1978, abandoned, which is a continuation of Ser. No. 658,694, Feb. 17, 1976, abandoned.

[51] Int. Cl.³ **H01F 7/02**
[52] U.S. Cl. **335/306**; 46/236; 335/302
[58] Field of Search 335/302, 306; 35/46; 46/236; 308/10

[56] **References Cited**

U.S. PATENT DOCUMENTS

2,323,837 6/1943 Neal .... 35/46 R

FOREIGN PATENT DOCUMENTS

642353 8/1950 United Kingdom 335/306

Primary Examiner—Harold Broome
Attorney, Agent, or Firm—Robert S. Smith

[57] **ABSTRACT**

A dish-shaped magnet in one form has an upper surface of a first polarity and a lower surface of a second polarity disposed in co-axial relationship to a second magnet having the opposite polar relationships. The magnetic fields in one form of the invention position the second magnet in spaced relation to the dish-shaped magnet. The apparatus has application as a novelty as well as for gyroscopic and other instrumentation apparatus wherein friction must be minimized. The upper magnet may be rotated either manually or by associated apparatus to provide gyroscopic stability.

13 Claims, 4 Drawing Figures

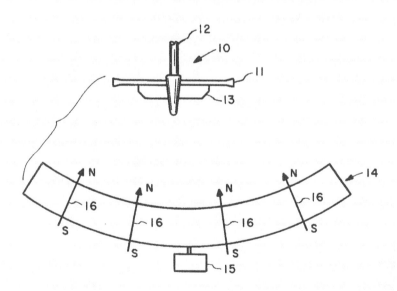

Figure 7. First page of Roy Harrigan's 1983 patent describing spin-stabilized magnetic levitation.

proved that neither of Harrigan's suggested forms of the base magnet was necessary.

In 1984, apparently unaware of the Harrigan patent, a Delaware inventor named Joseph Chieffo also reportedly succeeded in levitating a spinning magnetic top over a base magnet—in this case apparently a flat magnet. Chieffo's lawyer called his attention to the Harrigan patent, but advised him that his design was different enough that it would not infringe on Harrigan's patent. Chieffo placed a few classified ads offering his plans, and this led to others building and selling a few maglev tops generated by his designs. But sales were limited. We have found no documentation of how well maglev tops based on Chieffo's design actually operated.

The story of the Levitron itself began in 1993 when Bill Hones of Seattle, interested in marketing science toys, learned about Roy Harrigan's patent and went to Vermont to convince himself that Harrigan's device would actually levitate. Stories of this visit and its immediate aftermath vary, but the net result was that Bill's company, Fascinations Inc., was within a few months offering Chinese-built Levitrons for sale. And in April 1995, about the time I first encountered a Levitron in a science store, Bill and his father Ed Hones were granted U.S. Patent 5,404,062 for a "magnetic levitation device and method." Their patent cites Harrigan's patent as part of "prior art," and their new levitation device "comprises a first magnet with a polygonal, preferably square, periphery and a substantially planar upper surface magnetized normal thereto and a second magnet with an apparatus to rotate or spin the same." There were other differences in the many claims in Harrigan's and Hones's patents, but Harrigan stressed the use of a round dish-shaped base magnet, while Hones stressed a flat square-base magnet. The Hones analysis of magnetic fields and forces concludes, "other non-polygonal shapes, such as circular, elliptical, etc. do not appear to provide a region where both lifting and centering forces exist."

It is very difficult for a patent examiner to decide whether a new patent offers new features that make it clearly different from preexisting patents. In this case, the Hones patent emphasized the importance of the flat and square-base magnet, and offered a mathematical analysis that purported to show that a base magnet of circular or elliptical symmetry would not work. Their flat and square-base magnet was clearly different from Harrigan's dish-shaped circular magnet, and that difference and the supporting analysis probably influenced the examiner to decide that the Hones patent was sufficiently different from the Harrigan patent for it to be granted. However, their conclusion that a circular shape would not work is clearly wrong, since the base magnets of Levitrons today are circular. And several eminent scientists have severely criticized the mathematical analysis in the Hones patent, which seems to violate Earnshaw's rule.

Soon after the Levitron hit the market, it began to draw considerable interest in the scientific community. University professors who soon published papers in scientific journals or presented talks about the Levitron included Sir Michael Berry of Bristol University, England, Martin Simon of UCLA, Thomas Jones of the University of Rochester, Michael McBride of Yale, Ronald Edge of the University of South Carolina, and Rod Driver of the University of Rhode Island. Berry described the Levitron as "an ingenious mechanical device," and Simon and Jones deemed it "remarkable." All agreed that the levitating force resulted from the repulsive force between like poles of the base magnet and the top magnet, and that the rotation of the top was essential to the stability of the levitation. But they also argued that simply stating that gyroscopic action resists tipping of the top magnet is not sufficient to explain the stable levitation exhibited by the Levitron. It's more complicated than that and relates to the off-axis tilt of the magnetic field and how that affects the rotation of the top. Mathematical analyses of spinning magnets in magnetic fields are fairly complex, but yield results consistent with experimental observations of a limited height range of stability and with the Levitron's extreme sensitivity to the total weight of the magnet and to the rotation speed of the top. (When the rotation speed of Levitron tops decreases below about 20 turns per second, they become unstable, and both analyses and experiments indicate that they also become unstable above a maximum speed of rotation.) With regard to the vertical magnetic field produced by the base magnet along its central axis (the field that levitates the top), the equations for stability set specific criteria for the detailed variation of this field with height above the base. Theory indicates no need for square corners on the base magnet. In fact, in his published paper, Simon specifically demonstrated the levitation of a spinning top above a ring magnet of circular symmetry.

Several of the scientists who became fascinated with the Levitron and spoke or wrote about it are also among those who came to feel that the original inventor of spin-stabilized magnetic levitation, Roy Harrington, has been insufficiently recognized and rewarded by the marketers of the Levitron. Hones says that he originally offered Harrigan a contract to receive a modest percentage from the sales of the Levitron, but that they were unable to reach an agreement. On the Levitron website today, Harrigan is credited as the discoverer of "the effectiveness of spin in stabilizing a magnetically supported top such as the Levitron," but several of Hones's critics felt that in the early years of Levitron marketing, Harrigan was given insufficient credit. A 1996 article in *Discover* magazine discussing Hones and the Levitron says, "Only after six years of failure, when he was at the brink of despair, did he get the idea to try to levitate a magnetized spinning top." But there was no mention that Hones got the idea from Harrington. Among those feeling most strongly about this were Mike and Karen Sherlock, early retailers of the Levitron. In 1997, on their levitron.com website, they attacked Hones as essentially stealing the idea for the Levitron from Harrigan. This led to a legal battle between Hones and the Sherlocks, which in 1998 led to Professors Simon, Jones, and McBride presenting "expert reports" to the U.S. District Court in New Mexico challenging the Hones patent. They were countered by other magnetics experts supporting Hones. On Hones's side, Dr. Peter Campbell concluded, "the device disclosed in the Hones

Patent is distinct from, and in a commercial sense superior to, the device disclosed in the Harrigan Patent." Campbell's argument was based in part on the fact that a flat-base magnet was easier to manufacture than a dish-shaped magnet. On the other side, Simon attacked the scientific analyses in the Hones patent, concluding that the "essential 'teaching' of the Hones patent is totally wrong and has no commercial value." McBride's report criticizes both patents, stating, "Neither the Harrigan Patent nor the Hones Patent displays a correct understanding of the scientific basis for operation of the levitating top," but quotes Professor Berry of Bristol as saying, "The Levitron was invented by Roy Harrigan of Vermont, and developed and marketed by Bill Hones of Seattle."

None of the experts challenged the achievement of Hones in converting the idea first demonstrated by Harrigan into a marketable product. The question of the relative merits of the two patents never reached a judge. The legal case between Hones and the Sherlocks was eventually reduced to the simpler matter of trademark infringement, and the Sherlocks were required to relinquish their rights to the Web address "levitron.com." But the Sherlocks' "Hidden History of the Levitron" attacking Hones can still be found on the Web, not that well hidden. The Levitron may be "only a toy," but many mature adults developed very strong feelings about it.

The Super

A few years after the appearance of the Levitron, Bill Hones and Fascinations brought out the Super Levitron, which "FLOATS TWICE AS HIGH." As explained on the box, "This advancement is made possible through the use of very recently developed powerful neodymium iron boron magnets . . . Amazing to See Actually Floating Free in Space With No Strings Attached. Challenging to Operate, but Suitable For Ages 8 and Up." I'm very far into the "and Up" category, but agree that it, like the original Levitron, is "challenging to operate." But I finally could operate it, and it did indeed float about "twice as high." Next, as a materials scientist, I found it irresistible to study and compare the magnets in the original Levitron and in the Super Levitron.

The magnets in the base and top of the original Levitron were ferrite magnets. The ring magnet in the top was magnetized through its thickness like the upper magnet in Figure 3, with a south pole above and a north pole underneath. The base magnet was a square magnet 4 inches (10 cm) on a side and 0.5 inch thick, but its pattern of magnetization was more complicated, which I first revealed by applying a *magnetic viewing film* to its upper surface.

Magnetic viewing films are plastic films containing fine magnetic flakes that, like a compass needle, orient with the local magnetic field. If the local magnetic field is perpendicular to the film, the flakes turn perpendicular to the film and the surface appears dark. If the local field instead is parallel to the surface of the film, the flakes lie parallel to the surface, reflect more light, and the surface appears light. If the film is in contact with either a north pole or a south pole, the film appears dark. (By convention, we say that magnetic fields exit from north poles and enter into south poles. In either case, the local field is largely perpendicular to the pole surface.) However, in the boundaries between north and south poles, the field is largely parallel to the film, and the film appears light. For example, most refrigerator magnets are magnetized with alternating stripes of north and south poles. Put a magnetic viewing film on a refrigerator magnet and you see thick dark stripes separated by thin light stripes, the light stripes revealing the boundaries between north and south poles.

When I applied the magnetic viewing film to the upper or lower surface of the base magnet of the original Levitron, it revealed a white line, roughly circular, about 1.5 inches (3.8 cm) in diameter. Testing with other magnets and with a compass revealed that although the bulk of the upper surface of the base magnet was a north pole, the central region was a south pole. (On the lower surface, I observed the opposite—a central north pole surrounded by a south pole.) The base magnet clearly had a central cylinder magnetized in the opposite direction to the bulk of the magnet. When close to the base magnet, the top magnet was attracted to the central region but repelled by the surrounding region. I then held the base magnet vertical and felt the force between the base and top magnets in the horizontal direction so that I could separate the magnetic force from the downward force of gravity. By gradually increasing the separation between the top magnet and the center of the base magnet, I could feel the magnetic force gradually changing from attraction to repulsion. Clearly at the separation of roughly 1 inch where the spinning top levitates, the net force is repulsive, but it is attractive when the top is close to the base, where the spinning is initiated.

I then supplemented my observations with more quantitative measurements made with a *gaussmeter,* an instrument that measures the strength of magnetic fields (also sometimes called a teslameter or magnetometer). Most gaussmeters employ a "Hall sensor," named after Edwin H. Hall, who in 1879, while a graduate student at Johns Hopkins University, discovered that certain materials carrying an electric current in a perpendicular magnetic field develop a transverse voltage that can be used to measure the field. This phenomenon is called the Hall effect. (Not many scientists have their names permanently attached to important phenomena based on work they did as graduate students, but Hall was one of the lucky ones.) Typically the material used in a Hall sensor is a semiconductor like indium antimonide, which has a large Hall effect.

I used the Hall sensor to measure both the magnitude and the direction of the magnetic field at various points on the surface of the base magnet and above it, and the results were consistent with the qualitative observations made with the viewing film and with other magnets—that the center of the upper surface was a south pole and the surrounding region was a north pole. With the sensor in direct contact with the center of the upper surface, the magnetic field strength was about 0.05 tesla or 500 gauss, about 1,000 times the earth's magnetic field. (The standard international unit for magnetic field strength today is the tesla, after Nikola Tesla, a Croatian engineer and inventor who made many contributions to the electrical industry. However, also still in use is the gauss, named after Karl Friedrich Gauss, German mathematician. One tesla is equivalent to 10,000 gauss.) As I then raised the sensor carefully above the center, the vertical magnetic field first decreased in magnitude and then changed sign, changing from attractive to repulsive for a north pole, as observed qualitatively with the top magnet. Near the levitation height, the vertical field was about 100 gauss or 0.01 tesla. Here I should probably remind you that the upward magnetic force on the top magnet, the force producing levitation, depends on how rapidly the vertical magnetic field changes with distance—on its *gradient,* for example, in teslas per meter.

I then turned to the early Super Levitron, which had a square-base magnet of similar external dimensions. Now the magnetic viewing film revealed a central reversed-pole region more than 2 inches in diameter—much larger than in the original Levitron. The gauss-meter revealed, not surprisingly, that the magnetic fields emanating from both the top magnet and the base magnet were several times the strength of the fields of the magnets in the original Levitron. And when I measured the vertical magnetic field above the center of the base plate, it changed sign at a much greater height than in the original Levitron, presumably because the central region of reverse polarity was larger. It seems that the Super Levitron "floats twice as high" both because it has stronger magnets and because the central region of opposite polarity is larger. The larger central region of reversed polarity raises the height at which the necessary conditions for stability (as established mathematically by Simon and others) are reached, and the stronger magnets ensure that at that increased height the upward repulsive magnetic force is large enough to balance the downward gravitational force.

Although the Levitron patents do not clearly state that the base magnet should contain a central region of reversed polarity, they come close. The Harrigan patent says, "in other forms of this apparatus each magnet may have a peripheral region of one polarity and central region of the opposite polarity." The 1995 Hones patent states, "the height at which the dipole magnet levitates can be increased by weakening the magnetic field at the geometric center of the base magnet," and suggests achieving that with a central hole or by mounting a small disc magnet of opposite polarity. A later Hones patent, granted in 2003, offers another option: "applying a strong magnet field of opposite polarity to permanently weaken or partially demagnetize a portion of the central region." Whereas the first Hones patent emphasized the importance of a square or polygonal shape of the base magnet, their second patent argues that the "outer periphery of the shell or base magnet need not be polygonal in shape if the magnetic field of the base magnet is made nonuniform by partial demagnetization in a central region of the base magnet."

The Levitrons offered today probably all contain neodymium permanent magnets, but the name "Super Levitron" has disappeared. The three variations offered today on the Levitron website are the Omega, the Platinum Pro, and the CherryWood (Figure 8) The shims that were provided with the earlier Levitrons to allow leveling the base have been replaced with adjustable legs. All three of the new Levitrons are circular in symmetry; the square corners stressed in the first Hones patent are gone. And the CherryWood has a central hole above which the top levitates. The "weakening" of the vertical field at the center of the base magnet that was accomplished in the earlier Levitrons with a region of reverse magnetization is here produced by a hole. With a central hole, if the upper surface of the base magnet is a north pole, some of the magnetic field exiting from that surface curves back into the hole to reach the south pole on the under surface. Just as with the earlier Levitrons, the vertical magnetic field at the center of the base magnet is opposite in direction to the field that levitates the top.

Also offered among the Levitron products are an instructional video, a "Starter" to assist in spinning the top for those like me who are "not-so-nimble fingered," and a "Perpetuator" to maintain the spinning of the top for much longer periods than a few minutes.

Figure 8. CherryWood Levitron, with top magnet spinning and levitating above ring-shaped base magnet.

Without the Perpetuator, air resistance or other disturbing forces slow the rotation of the top until it no longer is stable and it falls. The Perpetuator is placed under the Levitron and provides synchronized magnetic forces to keep the top going for days or weeks or more instead of minutes. Professor Ken Libbrecht of Caltech, on his Web page devoted to science toys, details his attempts to use the Perpetuator to keep his Levitron running for record times. He records all the things that have interfered with the long-term spinning of his spinning top, including power outages, a web spun by a black widow spider, and an earthquake (perhaps he should move from California?). His site records a record of 315 days, but on another website, Nathan Tarler of Florida (Figure 9) reports a record of 586 days. He surrounded the Levitron and Perpetuator with a plastic cover to avoid disturbing

Figure 9. Levitron Master Nate Tarler with his Super Levitron, which spun and levitated continuously for a record 586 days.

air currents (and black widow spiders), and powers his Perpetuator with an Uninterruptable Power Supply to guard against power outages. He also has a copper puck below the spinning top, as electric "eddy currents" (see the next chapter) induced in the copper tend to dampen any oscillations of the top that might lead to its loss of stability. Perhaps by the time you are reading this book, Nathan or some other intense fan of the Levitron will have exceeded 586 days.

The importance of spin in stabilizing magnetic levitation of the Levitron is just one example of the intimate connection between spin and magnetism. The spin of the earth about its axis creates the earth's magnetic field. The spinning of magnets in generators, converting mechanical energy to electrical energy through electromagnetic induction (Fact 8—see next chapter), generates most of the electricity that powers our world. The spin of electric motors, using magnetic forces to convert electrical energy back into mechanical energy, does much of the work of the world. And the connection between spin and magnetism is also very intimate in the atomic and subatomic world. The spin of protons and neutrons is the basis of nuclear magnetic resonance imaging (MRI), the most important of the many uses of magnets in medicine. And the fundamental source of most of the magnetism of iron, and of ferrite and neodymium magnets and other magnetic materials, is the unbalanced spin of electrons. From electrons to Levitrons, spin is the source of much of the magic of magnets.

Inducing Uplift

Mutual Induction

As noted in Chapter 2, it was in 1820 that Hans Christian Oersted first demonstrated that an electric current, that is, electric charges in motion, produced a magnetic field. If the current was carried down a straight wire, the magnetic field produced was directed in a circle around the wire. No magnetic "poles" there. But if many turns of wire are wound together into a coil, the magnetic fields from each turn of wire add together, and the coil becomes an electromagnet with a north pole at one end of the coil and a south pole at the other (Fact 6 and Figure 2). About a decade later, Michael Faraday discovered the opposite effect—that the relative motion of a magnet and a conductor induced an electric current in the conductor (Fact 8). This is the important physical phenomenon known as *electromagnetic induction*. Electric charges in motion (current) produce magnetic fields, and magnetic fields in motion produce electric fields (which produce currents in a conductor).

Faraday attached a coil of wire to a galvanometer, a device to measure electric current (Figure 10). He found that if he inserted a magnet into the coil, a current flowed in the coil, but only while the magnet was moving. (The figure shows a single loop of wire. Faraday instead used a multiloop coil of wire to enhance the effect.) When he removed the magnet from the coil, a temporary current flowed in the opposite direction. Current flowed only during the entry and exit of the magnet. The experiment inspired a poem by Herbert Mayo:

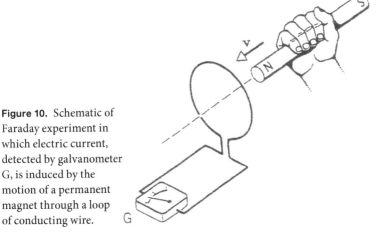

Figure 10. Schematic of Faraday experiment in which electric current, detected by galvanometer G, is induced by the motion of a permanent magnet through a loop of conducting wire.

> Around the magnet Faraday
> Was sure that Volta's lightnings play
> But how to draw them from the fire?
> He took a lesson from the heart:
> 'Tis when we meet, 'tis when we part
> Brings forth the electric fire.

While Faraday was exploring electromagnetic induction in the well-equipped London laboratory of Britain's Royal Society, an American named Joseph Henry was performing similar experiments in his own far humbler private laboratory, on his own funds, whenever he was not busy with his day job—spending seven hours a day to teach mathematics to young boys at the Albany Academy in Albany, New York (Figure 11). Although Faraday and Henry's experiments were seldom identical, they were very similar, and it has been argued that Henry actually discovered induction first. But he was not the first to publish; Faraday was. Nevertheless, it is fair to say that electromagnetic induction was the mutual and independent discovery of Faraday and Henry.

1799 JOSEPH·HENRY 1875
BAPTIZED·IN·THIS·CHURCH

Figure 11. Joseph Henry, co-discoverer of electromagnetic induction, teaching mathematics to boys at the Albany Academy as pictured in a window of the First Presbyterian Church of Albany, New York.

The two scientists separately continued to study the interplay between electricity and magnetism, and each made several important discoveries, often overlapping. For example, several sources credit *both* Faraday and Henry as constructing the first electric motor. Faraday actually constructed the first device to convert electrical energy into mechanical energy, the general definition of a motor, but Henry's first motor was closer to modern motor design than Faraday's. Henry also constructed the first powerful electromagnets and pioneered telegraphy. Unlike Faraday, Henry did his early research in relative obscurity. However, his research results eventually earned him recognition, at least in the United States, allowing him to leave his teaching post at Albany Academy and take a position at Yale. Later Henry became director of the Smithsonian Institute and president of the National Academy of Sciences, and is today recognized as the foremost American scientist of the nineteenth century. Most units used to express electrical and magnetic quantities, like volt, ampere, ohm, watt, tesla, gauss, and farad, were named for European scientists and engineers, but in 1893 the International Congress of Electricians named "henry" as the unit for inductance, making Henry the first American to have a basic scientific unit carry his name. Those so honored were still all Dead White Men, but at least they were no longer all Dead White European Men. It was one small step toward diversity.

But it is still Faraday's name that today is most closely associated with electromagnetic induction. It is Faraday's law of induction that was expressed by James Clerk Maxwell in what is today called the Faraday equation. One of Maxwell's four basic equations of electromagnetism, the Faraday equation states succinctly in just a few symbols that a time-varying magnetic field produces an *electric field proportional to the rate of change of the magnetic field.* And if that electric field is experienced by a conductor such as copper or aluminum, it produces currents.

It is rather ironic that Faraday's name is associated with a mathematical equation, since he intensely disliked formal mathematics.

He once wrote to André-Marie Ampère, "With regard to your theory [of electricity], it so soon becomes mathematical that it quickly gets beyond my reach." To Maxwell, he wrote, "When a mathematician engaged in investigating physical actions and results has arrived at his conclusions, may they not be expressed in common language as fully, clearly, and definitely as in mathematical formulae? If so, would it not be a great boon to such as I to express them so, translating them out of their hieroglyphics, that we also might work on them by experiment?" Faraday, generally considered to be one of the greatest experimental scientists who ever lived, was unable to appreciate the "hieroglyphics" and compact beauty of Maxwell's equations, which many scientists today consider to be the greatest intellectual achievement of the nineteenth century.

Maxwell's equations integrated the experimental and theoretical results of Oersted, Faraday, Ampère, Gauss, and others into a simple, unified form that, among other things, showed that they led to electromagnetic waves that traveled at the speed of light. (Many T-shirts worn by students at MIT preface a display of Maxwell's four equations with "And God said:" and follow them with "then there was light!") When I first lectured on Maxwell's equations at MIT, I extolled both their importance and their beauty, pointing to them on the blackboard and declaring, "Beauty is truth, truth beauty–that is all ye know on earth, and all ye need to know," citing it as from the poem Ode on a Grecian Urn by Shelley. After my lecture, a girl student who seldom spoke in class shyly approached me and said, "that poem was by Keats, not Shelley." Embarrassed, I admitted that I often got the two Romantic poets mixed up, and not only corrected myself at the start of the next lecture, but also added a relevant verse of my own that ended, "Jim Maxwell's rules impact it all, from quasar huge to atom small, but Keats and Shelley, you'll agree, can impact only you and me." For some reason, that incident inspired me to compose more verse and song lyrics for that course, which can be found online at PhysicsSongs.org. Some (but by no means all!) of my students seemed to enjoy them.

Another name closely associated with electromagnetic induction is that of Russian physicist Heinrich Lenz. His statement that an induced current flows in the direction that *opposes* the change that produces it has become known as "Lenz's law." (Lenz didn't get a unit named after him, but he did get a law. Ohm and Faraday got both.) Let's suppose that Faraday started to insert the north pole of a magnet into one end of a coil. Then the induced current flows in a direction that produces a north pole at that end of the coil and a repulsive force on the magnet, *resisting its insertion* into the coil. When Faraday withdraws the north pole of the magnet from the coil, the induced current flows in the opposite direction, producing a south pole of the coil that attracts the north pole of the magnet, *resisting its removal.* Induced currents are conservative—they *resist change.*

The device most often used to demonstrate Lenz's law in physics classes is the flux tube, in which a magnet is allowed to drop through a copper tube. The copper just below the falling magnet experiences an increasing magnetic field and the copper just above the falling magnet experiences a decreasing magnetic field. At each end of the falling magnet, the time-varying magnetic field results in a circular electric field and induced circular currents around the copper tube. And at each end, dutifully obeying Lenz, the direction of the current induced in the copper *resists change,* that is, slows the magnet's rate of fall. In the copper immediately below the falling magnet, the induced currents resist an increase in magnetic field and produce a magnetic field that pushes upward on the magnet (a repulsive force). In the copper immediately above the magnet, the induced currents resist a decrease in magnetic field and produce a magnetic field that pulls upward on the falling magnet (an attractive force). As a result, it takes several seconds for the magnet to fall through the tube, whereas another object of similar shape and weight that is not a magnet falls through the tube in less than a second. Through Faraday's law of induction and Lenz's law, the induced currents in the copper produce an upward (antigravity) force on the falling magnet.

The amount of current induced in the copper tube by the falling magnet depends on the *electrical resistance* of the copper, that is, on the ratio of voltage to current. (By Ohm's law, current in an electrical conductor is proportional to voltage.) The induced electric field and voltage depend on the rate of change of magnetic field, and a given electric field will yield more current if you lower the electrical resistance, which you can do by cooling the copper down, say with dry ice or, much better, with liquid nitrogen. The increase in induced currents produced by cooling the copper leads to an increase in the induced magnetic fields that resist the fall, and produce a much slower fall of the magnet through the copper tube.

The currents induced by time-varying magnetic fields are commonly called *eddy currents*. (They are *not* named for a scientist named Eddy. Induced currents are called eddy currents because they often flow in circular patterns, like whirlpools or eddies in water.) And the slowed falling of the magnet through the copper flux tube is a simple example of a much-used phenomenon, "eddy-current braking." In some of the "free fall" towers in amusement parks mentioned earlier, after a few seconds of free fall, cars are thoughtfully decelerated by magnets on the falling car moving past stationary structures of copper (i.e., by eddy-current braking). Compared to usual friction braking, eddy-current braking requires no direct physical contact and associated wear, and, if permanent magnets are used, there is no power consumption. Eddy-current braking is frequently used to slow roller coasters, high-speed trains, and rotating machinery. Eddy-current braking can occur if the stationary part is a conductor and the moving part is a magnet, as in the flux tube, or if the magnet is stationary and the conductor is moving. It's their *relative motion* that counts. Many coin machines contain stationary magnets past which the coins fall, and the time of fall is determined by eddy-current braking and therefore by the electrical resistance of the coin, allowing the machine to detect false coins.

So electromagnetic induction and eddy-current braking can produce antigravity forces to slow falling objects, but how about levitation? One simple experiment can at least show that eddy currents can produce lift. Place a ring of copper on a table, insert a powerful neodymium magnet into the ring, and then quickly lift the magnet. That will induce eddy currents in the ring that *resist change* and produce an *attractive* force between the magnet and the copper, and that force will lift the copper ring briefly off the table. The eddy currents induced in the copper quickly decrease because of the electrical resistance of copper, and the ring instantly falls back down. But at least this time the antigravity forces from the eddy currents produced some temporary upward motion instead of just a slowing of a falling magnet. If you decrease the electrical resistance of copper by cooling it to very low temperatures, say with liquid nitrogen, the eddy currents induced by raising the magnet will last longer, and the ring can be lifted a bit longer and a bit higher. What if the ring had no electrical resistance at all? We'll consider that case in Chapter 7. There the induced currents and the resulting lift by attractive forces might last a very long time indeed.

It is possible to produce *sustained* lift from eddy currents if the relative motion between the magnet and the conductor is sustained. One popular demonstration of lift produced by relative motion of a magnet and a conductor employs a rotating copper sheet, say placed on a phonograph turntable, and a powerful magnet placed on the tone arm where the phonograph needle usually is. (For readers too young to know what a phonograph is, ask your parents.) When the turntable is stationary, the magnet rests lightly on the copper sheet. But when the turntable rotates rapidly enough, the magnet lifts a few millimeters (about an eighth of an inch) above the sheet. The eddy currents induced in the copper under the magnet by relative motion make that part of the copper sheet an electromagnet that, following the dictates of Mr. Lenz, opposes the field of the permanent magnet, repels it, and lifts it off the sheet.

How about full *contact-free* levitation from eddy currents induced by relative motion? An eddy current levitator developed and patented by Andrew Alcon in 1994 can be seen in a UCLA website dealing with physics demonstrations. He spins two adjacent cylindrical aluminum rotors in opposite directions and levitates a magnet above the center line between the two spinning rotors. The high-speed relative motion of the aluminum rotors with respect to the magnet induces currents in the aluminum that create a magnetic field that repels the magnet and levitates it. The groove formed between the two rotors keeps the magnet stable against sideways motions. A similar device built by Bill Beaty of the University of Washington, also visible online, uses two parallel thin-walled copper tubes about a foot long and a little over an inch in diameter spun at high speed in opposite directions. As Beaty pointed out, this is a dangerous experiment with opportunities for serious accidents and added, "don't try this at home." But the Alcon and Beaty devices are conceptually simple and one of the ways to get full contact-free levitation with relative motion alone. We shall see in a later chapter that several approaches to maglev trains use eddy currents produced by sustained relative motion between magnets and conductors to produce levitation. Such systems are called electrodynamic.

Flinging Rings and Things

In the cases considered above, the time-varying magnetic field inducing currents in a conductor by Faraday's and Lenz's laws was produced by the relative motion of a permanent magnet and a conductor. However, since magnetic fields can be produced by electric currents (as we learned from Oersted and Fact 6), changing magnetic fields can also be produced by *changing electric currents,* with no need for either a permanent magnet or relative motion of the conductor. An experiment usually attributed to Faraday involves a doughnut-shaped piece of iron and two coils of wire wound around separate sections of the iron (Figure 12). If a switch S is thrown that suddenly connects one coil to a battery V, a brief pulse of current is observed in the second coil. What has happened is that closing the

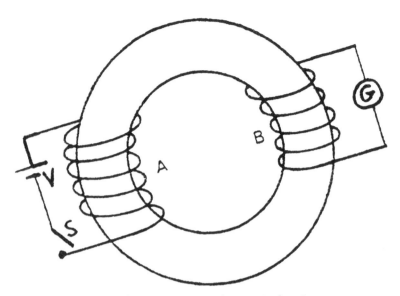

Figure 12. Schematic of electromagnetic induction via changing currents. When the switch S is closed, voltage V produces a current in coil A that magnetizes the doughnut-shaped piece of iron. The increase in magnetization of the iron induces a pulse of current in coil B that is detected by galvanometer G. When the switch is later opened, current stops flowing in coil A, the iron loses its magnetization, and that change induces a current pulse in the opposite direction in coil B. Current flows in coil B only when the magnetization of the iron changes, that is, only when switch S is closed or when it is opened.

switch produces a sudden current in the first coil that suddenly magnetizes the iron doughnut. That produces a sudden change in the magnetic field felt by the second coil and induces a current. Through Lenz's law, the direction of the induced current produces a magnetic field opposing the increasing field in the iron. Once a steady current is flowing in the first coil, no further change of magnetic field takes place and there is no induced current in the second coil. When the switch is later opened, current ceases to flow in the first coil, the iron becomes demagnetized, and the sudden decrease in magnetic field induces a current in the second coil in the opposite direction, in the direction that produces a magnetic field in the direction of the decreasing magnetic field of the iron—it *resists change.*

In the earlier experiment with the moving magnet (Figure 10), current was induced in one direction when the magnet entered the coil and in the opposite direction when the magnet exited the coil. It was magnetic *change*—"when we meet" and "when we part"—that induced current. The same is true in this latest experiment. Current was induced in the second coil only when the switch was closed and when it was opened—when *change* in the current to the first coil produced *change* in the magnetic field felt by the second coil.

The term *mutual induction* used in the previous section to refer to the mutual discovery of induction by Faraday and Henry is more often used to describe situations like that in Figure 12, in which one electric circuit is influenced by changes in another electric circuit. This is the basis of *transformers,* in which different numbers of turns in two coils of wire linked by iron can produce desired increases or decreases in voltage. The same effect of currents induced in one coil by changes in another can also be observed in two coils in close proximity without the iron, but the presence of iron greatly increases the magnetic fields involved (Fact 7) and enhances the effect.

A related popular lecture demonstration is the "jumping ring" experiment, sometimes called the ring fling. Here when the switch is closed and a sudden current flows in the coil to produce a sudden magnetization of a vertical iron rod, the changing magnetic field induces a large current in an aluminum ring around the rod. Here the aluminum ring is like the second coil in the earlier experiment. (The effect can also be produced with a ring of copper or any other good conductor, but aluminum has the advantage of being light and easy to lift.) By Lenz's law, the induced current in the aluminum ring produces a field opposing the field of the coil and the iron, and the result is a repulsive force that sends the ring flying several feet up into the air. (Most instructions for the experiment warn that you should make sure that there are no ceiling lights directly above the experiment.) If the aluminum ring is cooled with liquid nitrogen to decrease its electrical resistance, more current is induced and it jumps even higher. But if a ring is used that has a small gap cut in it so that induced currents cannot flow completely around the ring, the ring does not jump.

The "ring fling" has been a popular lecture demonstration for many years in freshman courses in electricity and magnetism (E&M in academic lingo). Many thousands of college freshmen, and some high school students as well, have seen electromagnetic induction and Lenz's law demonstrated by flinging rings. And in the Greater Boston area, many thousands of elementary school children have seen a fuzzy stuffed animal, often Garfield the cat, flying through the air high above their heads in their school auditoriums, thanks to Paul Thomas, MIT's "Mr. Magnet." For seventeen years, Paul traveled to area schools with an hour-long show featuring many demonstrations of the wonders of electricity and magnetism, including the "Boomer" developed by MIT Professor Harold Edgerton. In the Boomer, a short pulse (about a millisecond) of high current (about 1,000 amperes) is delivered from a bank of capacitors (charge-storage devices) to an electromagnet coil. On top of the coil is a tethered aluminum plate, on which Paul places Garfield before initiating a 5-4-3-2-1 countdown and pushing the button that unleashes the current. Accompanied by a very loud noise (hence the name Boomer), the pulse of current produces a pulse of magnetic field above the coil, which induces strong eddy currents in the aluminum plate, producing a powerful repulsive upward force. The plate is firmly tethered with a strong strap attached to one side, and so it simply flips over with the pulse. Garfield, however, is not tethered. He is pushed up by the plate and flies away, to the delight of the school children (most of whom also enjoy the boom).

In another demonstration of upward magnetic forces produced by the Boomer, Paul places a thick copper plate above the coil and weighs down the center of it with a heavy lead disc. The lead is heavy enough that it holds down the center of the copper plate when the

Boomer is activated, but the eddy currents and resulting upward forces on the outer regions of the copper plate bend the plate into a bowl, demonstrating the huge forces that induction can produce. This is a simple example of the industrial process known as electromagnetic forming or magneforming.

A battery produces DC or direct current, which flows in only one direction, but the electric current delivered from the outlets in your house instead alternates in direction, continually flowing back and forth, and is called *alternating current* or AC. Alternating currents of course produce alternating fields, a form of time-varying magnetic fields. If in the "ring fling" experiment the coil carries alternating currents, they produce alternating magnetization of the iron rod, which in turn induces alternating currents in a surrounding aluminum ring. In this case, the input current can be adjusted to maintain the ring at a fixed height, the upward repulsive magnetic force producing a form of magnetic levitation rather than an upward moving projectile. But the ring is constrained by the iron rod, much as the ring magnet was constrained by the pencil in Figure 3, so this is not fully contact-free levitation.

Nevertheless, full contact-free magnetic levitation by alternating currents induced by alternating magnetic fields is indeed possible and can be seen in Figure 13. Here you see two school children fascinated by the levitation of an aluminum frying pan by Paul Thomas. An AC electromagnet is hidden within a large wooden box and can be turned on and off by a switch hidden in Paul's hand. The alternating currents in the electromagnet produce alternating magnetic fields above the box that induce alternating electric fields and currents in the frying pan. By Lenz's law, the field produced by the induced currents in the pan opposes the field from the electromagnet, resulting in a repulsive force and levitation. The electromagnet is shaped to produce a magnetic field that also produces sideways restoring forces that resist sideways motion of the pan. Stable contact-free levitation here does not violate Earnshaw's rule because Earnshaw considered only static fields and magnets, not AC fields and in-

Figure 13. Paul Thomas of MIT amazing two Cambridge school kids by magnetically levitating an aluminum pan. Within the box underneath the pan is a large AC electromagnet, which induces eddy currents in the pan and creates an upward repulsive force on it.

duced currents. AC eddy currents are capable not only of levitating conductors above electromagnets, as in Figure 13, but also of doing the reverse—levitating an AC electromagnet above a conducting plate. On YouTube, you can see videos of an AC coil levitating above an aluminum plate.

The levitation of the frying pan would appear less magical if Paul instead were to turn on the electromagnet by visibly closing a switch on an electrical panel instead of with a switch hidden in his hand. Here the magical effect on his audience combines the magic of electromagnetic induction with the magic of wireless communication. (If he wanted to, Paul could even enhance the effect for Harry Potter readers by waving a magic wand and chanting "Wingardium

Leviosa!" when he secretly pushes the hidden switch that turns on the AC electromagnet, but I don't think he's gone that far.)

Eddy-Current Heating and Levitation Melting

When Paul Thomas levitates that aluminum frying pan, the pan gets hot pretty quickly from resistive heating by the eddy currents. If Paul wanted to fry eggs, the pan quickly gets hot enough to accomplish that. The alternating current that we get from the electric outlets in our houses is alternating at only 60 cycles per second—that is, it goes from maximum current in one direction to maximum current in the opposite direction and back again in one-sixtieth of a second, and that's the frequency of the magnetic fields that heat and levitate Paul's aluminum pan. But many applications of eddy-current heating and levitation, including Edgerton's Boomer, involve AC fields that change much faster than that—at several thousand cycles per second. The unit we use to define AC frequency in cycles per second is the hertz (usually abbreviated as Hz), named after German physicist Heinrich Hertz, who in 1886 first clearly demonstrated the existence of electromagnetic radio waves. (Hertz had noted that the electromagnetic waves predicted by Maxwell's equations could occur at frequencies other than light frequencies.) Although Paul's pan was levitated and heated by 60 Hz fields, most applications of eddy-current heating, including induction furnaces used in industry for heating and induction cookers used in homes to cook foods, use frequencies of many thousand hertz, many kilohertz (kHz). The Faraday equation tells us that the electric field induced by changing magnetic fields is proportional to the rate of change of the magnetic fields. High-frequency AC fields change faster than low-frequency fields, and thus induce higher eddy currents and produce more heating.

There are many other examples of high-frequency AC fields being used to produce eddy-current heating and levitation. With different configurations and arrangements of AC coils, flat circular or rectangular conducting plates, hollow or solid metallic spheres and cylinders, and liquid metal droplets have been stably levitated.

Usually levitation via eddy currents is done with various arrangements of coils under the levitated metal object, but levitation is also possible with a pair of matched AC coils, one above and one below. A hollow aluminum sphere has been stably levitated between two identical coils carrying currents in opposite directions so that the magnetic field is zero at the center point between the coils and increases in every direction outward. That produces both horizontal and vertical restoring forces.

One of the most fascinating examples of levitation with AC electromagnets is the process called *levitation melting,* in which high-frequency eddy currents are used to heat, melt, and levitate metals. Although levitation melting has been studied for a wide variety of metals, it is especially useful for highly reactive metals such as titanium or niobium that would be easily contaminated by chemical reactions if they were melted in contact with a crucible, and offers a means of containerless processing of such metals. The eddy-current levitation melting of reactive metals is done in vacuum or in inert gases such as argon. The electromagnetic forces tend to stir the liquid, so encourage uniform distribution of alloying elements. Metal samples as heavy as 1 kilogram in weight (2.2 pounds) have been levitation melted.

The windings for levitation melting usually involve a lower conical set of coils within which the metal is levitated and one or more upper "bucking coils" in which the currents flow in the opposite direction and provide a local vertical magnetic field in the opposite direction. The combination of the lower and upper coils provides along the coil axis a vertical magnetic field that is strongest near the narrow bottom of the conical coil, decreases with height within the widening conical coil, goes through zero, and reverses within the upper coil. That vertical field, alternating several thousand times a second, induces most of the eddy currents and provides levitation and heating first to the solid metal and, once the melting point is exceeded, to the liquid metal. Changes in the magnetic field for positions off the coil axis provide sideways forces that provide horizontal stability to the heated and levitated metal. The coils are usually water-cooled copper tubes (they don't want those to melt!), and the AC frequencies used are usually of the order of 10 kHz (10,000 cycles per second). Much trial and error was necessary over the years to determine optimum coil arrangements and optimum power levels and frequencies to achieve the desired temperatures for different levitated materials, in part because the heating of each material depends on its electrical resistance, which is different for different materials and changes with temperature and with melting. And the same eddy currents that are responsible for the levitation are responsible for the heating, so achieving specific desired temperatures in the levitated melt is a challenge. Variations of density and weight must also be taken into account.

Electromagnetic induction can do many things. It generates electricity in our power plants, changes voltages and currents in transformers, moves induction motors (including the linear induction motors that propel some maglev trains), and provides eddy-current braking, damping, heating, and levitation, including the levitation of maglev trains at record-breaking speeds. In a wide variety of sensors used in industry, electromagnetic induction is used to measure position, speed, and many other important quantities. And it does lots more. Much of modern technology is enabled by electromagnetic induction, the phenomenon discovered mutually and independently in the 1830s by Michael Faraday and Joseph Henry.

Fields of Force

Earlier in this book, when discussing Fact 1 of the basic "facts about the force," I mentioned the *magnetic field* of the earth. I mentioned an *electric field* near the end of Chapter 1 and the *gravitational field* of the earth in Chapter 2. There were lots of electric fields and magnetic fields throughout the present chapter. (If I had been referring to Stephin Merritt's popular band The Magnetic Fields, I would have used capital letters.) I suspect that most readers have heard about magnetic fields and electric fields before, but this might be a good place to clarify them a bit, since this chapter is where Faraday and Maxwell made a strong appearance, and they are often credited with helping to develop the modern concept of fields.

If you look up "field" in your dictionary or Wikipedia, you'll find lots of meanings. There are football fields and Strawberry Fields, fields of candidates, fields of view, fields of study, and fields of endeavor. (I mentioned Sally *Field* in Chapter 1; she's in the acting field.) The word "field" has different meanings in different fields.

In physics, invisible magnetic, electric, and gravitational *fields* are concepts that represent the *forces* exerted on one thing at one place by other things at another place. Things physically separate exert an influence on each other in the form of *force at a distance.* Everything is connected to everything else. You see the gravitational *field* of the earth through the results of the *force* the earth exerts on Newton's apple—you see the apple fall. You can *measure* that force if you want; instead of letting the apple fall, you can put the apple on a scale. And you can use the mathematical theory developed by Newton to *calculate* the gravitational force of the earth on things like apples and the moon, and the effects of those forces on their motion. The apple may fall down and hit you on the head, but the moon won't (thank goodness) because its orbital velocity keeps it falling *around* the earth (like those astronauts in the orbiting space shuttle we discussed earlier)

Like invisible gravitational fields, invisible electric and magnetic fields at any positions in space can be measured, that is, electric and magnetic *forces* can be measured, by putting charged particles, current-carrying wires, or magnets in those positions. And with well-established mathematical theories, you can calculate the fields and forces produced there by distant charges, currents, or magnets. Sometimes fairly simple equations will be sufficient, and sometimes we may need a computer to do the calculations, but the results can be used to understand how things work and to design new things that may work better. So although you can't see all those "fields" with your eyes, with the help of mathematics and physical theories we can actually use the concept of fields to understand and explain things, and to design things. The concept of a field to define force at a distance has proved to be a very useful concept.

One other point—forces have both strength and direction. So magnetic fields that represent magnetic forces also have both strength and direction. (Mathematicians and physicists call such things vector fields.) Magnetic fields generally curve through space like the field lines we drew in Figure 2, and their directions in space around a magnet can be revealed, as you may remember from school, by sprinkling iron filings on a piece of paper placed over the magnet. A magnetic field may point in any direction, but sometimes the important part of it is the part or *component* of the field (force) in a particular direction, say the vertical or horizontal component. An ordinary compass rotates freely about a vertical axis and shows us the orientation of the horizontal component of the earth's magnetic field—usually roughly north–south. To see the vertical component, you'll have to get a different magnetic device specifically designed to rotate freely about a horizontal axis; it's called a dip needle. If you're at the magnetic north pole of the earth (which nowadays is among the islands of northern Canada), your ordinary compass will be useless, but a dip needle will point straight down because here the magnetic field is entirely vertical, with no horizontal component. (It is the vertical component of the earth's field that led some bacteria to evolve with internal compasses to keep them away from the toxic oxygen-rich layers atop the mud they live in, as discussed in *Driving Force*.) At the magnetic equator of the earth, the field will be entirely horizontal, with no vertical component.

Now let's leave our basic discussion of fields, travel back to whatever latitude you are currently living in (where the earth's magnetic field probably has both horizontal and vertical components), and return to applying the concepts of magnetic fields and forces to our main topic—the field of magnetic levitation. We turn next to levitation that does not require relative motion or time-changing fields and instead requires only a kind of material that Reverend Earnshaw did not consider—a *diamagnetic* material.

Flying Frogs

The Ig Nobel Prize

Winning an Ig Nobel Prize is not nearly as prestigious or as financially advantageous as winning a Nobel Prize, but it often gets as much press coverage. ("Ig Nobel" is pronounced with emphasis on the "bel," supposedly to avoid the suggestion that there is anything "ignoble" about it.) Initiated at MIT in 1991, the annual October ceremony for granting these satirical awards moved to Harvard a few years later and continues there today. At the Ig Nobels, unlike the formal Nobel ceremonies in Sweden, tuxedos are not required, and you do not get to meet and shake hands with a king. However, there are usually a few real Nobel laureates participating in the event to give it the illusion of class—and perhaps to demonstrate that they have a sense of humor.

Magnetic levitation was finally recognized by the Ig Nobel awards committee in 2000, when Andre Geim of the University of Nijmegen, the Netherlands, and Michael Berry of Bristol University, England, were awarded the Ig Nobel Prize in Physics "for using magnets to levitate a frog and a sumo wrestler." Here the committee mistakenly gave credit where credit was not due, since Geim and Berry had not really used magnets to levitate a sumo wrestler. That feat was accomplished by a Japanese laboratory team with the help of superconductors and will be discussed in the next chapter. The awards committee had apparently been misled by the Web page of the University of Nijmegen High Field Magnet Laboratory entitled

"The Real Levitation," which includes a picture of a levitated sumo wrestler. But the caption under the levitated wrestler now clearly states "this is not our work." Records have been corrected, and in most online listings of past Ig Nobel Prize winners, the reason for the 2000 award to Geim and Berry has been curtailed to "for using magnets to levitate a frog." Reason enough for an Ig Nobel Prize.

Not all winners of Ig Nobel Prizes are pleased. Some consider the awards, which can be viewed as ridiculing legitimate scientific research, as not only *ig*noble, but also *ig*norant and even *ig*nominious. Consider some of the other awards in 2000. The Ig Nobel Prize in Biology was for a study of "the comparative palatability of some dry-season tadpoles from Costa Rica" (graduate students eating tadpoles to test whether unpleasant taste was a successful defense mechanism against predators), and the Ig Nobel for Medicine was for a Dutch MRI study of "male and female genitals during coitus and female sexual arousal" (the prize recipient labeled his work "Love Among the Magnets"). With these two awards and the Physics Prize, both frogs and magnets got double recognition that year. A group from Scotland received the 2000 Ig Nobel Prize for Public Health for their alarming report, "The Collapse of Toilets in Glasgow," and the Peace Prize went to the British Royal Navy, for ordering its sailors to stop using live cannon shells in practice and to instead just shout "Bang!" Questionable company for a scientist who takes his own research seriously.

Although some prize recipients with a limited sense of humor see only the *ig*nominy of the prize and choose to *ig*nore their award, Andre Geim and Michael Berry announced in a press release on "The Physics of Flying Frogs,"

> We are pleased to accept the Ig Prize because we have always considered it a duty to make physics more understandable and bring it closer to nonscientists. We think the prize acknowledges our contribution in that direction. Although some people tend to judge the quality of science by the seriousness of the researchers doing it, there

are lots of examples where good science has been fun—and science does not have to be boring to be good . . . Let there be more science with a smile!

Michael Berry is a theorist who helped to explain the conditions for stable levitation of the frog, but he made it clear in his accompanying statement (entitled "levitation without meditation") that "the flying frog was Andre Geim's experiment." Although Berry did not make the trip to Cambridge for the awards ceremony, Geim did, and I had the pleasure of having lunch with him that day at Harvard Square. He clearly had mixed feelings about the "honor" of receiving an Ig Nobel Prize, but he knew that his public announcement several years earlier of the magnetic levitation of frogs had generated great interest in the general public. Many schoolteachers and pupils had been stimulated by the topic and contacted him seeking information and advice. And Geim also heard from many practicing scientists who were surprised and fascinated by his demonstration that even the very weak magnetic properties of "nonmagnetic" materials like frogs could produce levitation in high-field electromagnets. He realized that even an intentionally humorous occasion like the awarding of Ig Nobel Prizes could help to enhance public appreciation of the wonders of science in general and magnetism in particular, and so he came to accept his prize. At the awards ceremony, Geim's acceptance speech was limited to only 30 seconds. But most accounts of the event mentioned the flying frog, and the publicity helped to further increase the awareness of scientists and nonscientists alike of the underlying physical phenomenon that allowed the frogs to fly, their *diamagnetism.*

Diamagnetism and the Moses Effect

Only a few materials like iron are strongly magnetic. Most materials, including frogs, are about a billion times less magnetic than iron, and their magnetic properties can be measured only with very sensitive equipment. Materials like iron that are strongly attracted to magnetic fields are called *ferromagnetic,* the prefix "ferro" coming from *ferrum,* the Latin word for iron. They derive their remarkable properties from the magnetism of their individual atoms, mostly produced by the unbalanced spins of electrons, plus a strong "exchange" interaction between the net spins of neighboring atoms that helps them to act in unison. (Some have characterized it as "electronic fascism.") Materials that are *weakly attracted* to magnetic fields are called *paramagnetic.* Most paramagnetic materials have atoms with unbalanced spins, but the net spins remain uncoupled from the net spins of neighboring atoms and tend to point in random directions (electronic anarchism?).

Materials that are *repelled* by magnetic fields are called *diamagnetic,* a term introduced by Faraday (of induction fame) in 1846. Water and most organic molecules, the major constituents of frogs and humans, are weakly diamagnetic. Application of magnetic fields to diamagnetic materials produces slight alterations in the paths of their electrons, and those altered motions of electrons yield a tiny net magnetic field *opposing* the applied field and proportional to it. There is a similar effect in paramagnetic and even ferromagnetic materials, but it is outweighed by the opposing and larger effect of the alignment of unbalanced electron spins with the applied magnetic field. Diamagnetic materials have no, or very few, unbalanced spins. Their electrons still spin, but occur in pairs that spin in opposite directions and cancel each other out.

Although inductive levitation, the topic of the previous chapter, and diamagnetic levitation, the topic of this chapter, both involve electron motions that produce opposing magnetic fields and repulsive forces, they are very different phenomena. Inductive levitation occurs only for time-varying magnetic fields applied to conductors and involves long-range electrical currents that can produce substantial heating. Diamagnetic levitation occurs with steady magnetic fields applied to *any* materials, including insulators, and involves only altered electron motion on the atomic or molecular level.

In diamagnetic materials, the ratio of the tiny opposing magnetic field to the applied external field that produced it (i.e., the *magnetic susceptibility* of the material) is negative and is typically only a few *millionths*, a few *parts per million* (ppm). The negative susceptibilities of water and most organic materials are very small, less than 10 ppm. Although the diamagnetism of water is a very weak effect, it can produce noticeable effects if the applied magnetic field is very large. The easiest way to demonstrate this is to place a neodymium magnet (which can produce a magnetic field at its poles of roughly one tesla) in a small dish with a pole facing upward and then pour water in the container until it is just slightly deeper than the height of the magnet. The thin layer of the water above the magnet pole will be weakly pushed away, and with proper lighting the slight depression of the water surface above the magnet can be seen.

A much more striking demonstration of the diamagnetism of water was made in 1994 by two Japanese researchers who placed a container of water within the central hole of a horizontal electromagnet about 8 inches (20 cm) long that produced at its center a very high magnetic field, up to 8 tesla or 80,000 gauss. Along the length of the electromagnet, the magnetic field is strongest at the center and falls off as you approach the ends. The water, which seeks low magnetic fields, is repelled horizontally in both directions from the high-field center toward the low-field ends. In their magnet, the water was repelled from the center strongly enough that it actually became nearly totally parted and the bottom of the container could be seen. This "parting of the waters" reminded the researchers of the parting of the Red Sea in the biblical story of Exodus and so they dubbed it the "Moses effect."

It is a remarkable effect, even though it is on a much smaller scale than the biblical account. Although the diamagnetism of water is very weak, the high magnetic field in the electromagnet, and the fact that the field decreased rapidly from the center of the magnet toward its ends, pushed the water strongly toward the ends of the magnet. Some years ago, I was inserted into a large electromagnet for an MRI of my lower back, and I'm glad that I wasn't thinking of the Moses effect that day. Otherwise, I might have worried that the magnetic fields, pushing the lower half of my body toward one end of the magnet and the upper half of my body toward the other end, might split me in two like the water in the Japanese experiment. Concern about the Moses effect acting on a patient in an MRI magnet presumably motivated General Electric Medical Systems, a major manufacturer and supplier of MRI magnets, to ask one of its engineers to consider the problem.

In our discussion in Chapter 3 of the forces between magnets, we noted that the net magnetic force on a magnet depended on the rate of change of magnetic field with distance, its *gradient*. However, when we deal with magnetic forces on diamagnets like water (or me), we must also include the fact that the magnetic strength of the water, weak as it is, is proportional to the strength of the applied magnetic field. Thus the horizontal magnetic repulsive force exerted on the water in the "Moses effect" experiment depends both on the strength of the magnetic field (tesla) *and* on the rate of change of magnetic field with distance along the electromagnet, that is, on the *gradient* of magnetic field (tesla per meter). (The same is true, as noted in Chapter 3, for the much stronger *attractive* force exerted by a magnet on iron.) The magnetic force in this case depends on the magnetic field *times* its gradient, tesla *times* tesla per meter, so the magnetic force depends on tesla *squared* per meter. For magnets of the same dimensions, reducing the magnetic field strength by a factor of four (e.g., from 8 tesla to 2 tesla) would therefore decrease the magnetic force by a factor of sixteen (four squared). The force would also depend on the length of the electromagnet. A longer magnet has a slower decrease in the magnetic field with distance from the center, and this decreased field gradient would further decrease the magnetic forces. Most MRI magnets in use are weaker and larger than the electromagnet used by the Japanese researchers. As a result, when the GE engineer reproduced the Japanese experiment within a typical GE MRI magnet, he observed a Moses effect that was much weaker than the one observed by the Japanese. And although we humans contain lots of water and other diamagnetic materials, our bones, muscles, skin, etc., serve to hold us together and keep us from splitting in two in response to the magnetic forces. The small magnetic forces pushing my upper and lower body parts toward opposite ends of the magnet that day may have made me a tiny bit taller while I was in the MRI magnet, but I came out in one piece, as have millions of other patients safely exposed to the high magnetic fields (and field gradients) of MRI magnets.

In the "Moses effect" experiment, both the magnetic field of the electromagnet and its gradient, and therefore the magnetic force, are horizontal and directed toward the ends of the electromagnet. Clearly, if the electromagnet had been vertical, the repulsive force on the water would have been vertical, which indicates the possibility of levitation. But the first demonstrations of the levitation of diamagnetic materials above the end of an electromagnet were made not on water, but on materials whose diamagnetism was much stronger than that of water.

Diamagnetic Levitation and Stabilization

The most strongly diamagnetic materials known are bismuth and graphite, with average susceptibilities typically about 170 ppm, or about twenty times the diamagnetism of water. This makes them easier to levitate than water and frogs. Actually, levitation of small pieces of bismuth and graphite above the end of a vertical electromagnet was demonstrated by a German researcher as early as 1939. In 1991, using a vertical electromagnet with a much stronger magnetic field (about 20 tesla), two French scientists reported the levitation above the magnet not only of bismuth and graphite, but also of wood, plastic, water, ethanol, and acetone. These early reports of diamagnetic levitation published in scientific journals drew little attention, even in the scientific community. It was the flying frogs of Andre Geim that captured the public imagination and led to further approaches to demonstrate diamagnetic levitation.

Graphite, the major component of pencil "lead," consists of carbon atoms linked to each other in a hexagonal planar network reminiscent of chicken-wire fencing or bathroom tiling. Each carbon atom in the network is strongly linked to three other carbon atoms in the planar layer, but the atoms in each plane are only very weakly linked to those in neighboring layers. As a result, electrons flow freely within the individual layers, but only with difficulty between the layers. In ordinary graphite, the planes are oriented in all directions, but in specially prepared "pyrolitic" graphite, most of the "chicken wire" planes of carbon atoms lie in one direction, parallel to each other. With this material, if a magnetic field is applied perpendicular to the planes, the resulting electron flow is around loops of atoms within the planes and produces a diamagnetic susceptibility of pyrolitic graphite as high as 450 ppm, much higher than that of ordinary graphite. (If the magnetic field is instead applied parallel to the planes, there is much less electron flow, and the magnetic susceptibility is *lower* than that of ordinary graphite.) Pyrolitic graphite, for a magnetic field perpendicular to the planes of linked carbon atoms, is currently the most diamagnetic material known (except for superconductors). This increased diamagnetism of oriented pyrolitic graphite facilitates various demonstrations of diamagnetic levitation.

One of the simplest such demonstrations is that of a thin piece of pyrolitic graphite levitated above an array of four neodymium magnets arranged with alternating north and south poles up (Figure 14). Although the height of levitation is only 1 or 2 millimeters, the setup is so simple that it can be easily passed around to students sitting around a table, who can experiment with it. Because the graphite is repelled by magnetic fields, it seeks a position with minimum vertical magnetic field, tends to align with the boundaries between magnets, and therefore is stable against sideways displacements. A set with the graphite sheet and four neodymium magnets is offered by several scientific supply houses.

There are other configurations of neodymium magnets and graphite with which you can demonstrate diamagnetic levitation.

Figure 14. Diamagnetic levitation of a thin flake of graphite above four neodymium permanent magnets.

For example, a ring magnet with a rod magnet magnetized in the opposite direction within the central hole (reminiscent of the pole arrangement of a Levitron base magnet) will levitate a pyrolitic graphite sheet. Two rows of neodymium magnet cubes aligned to form a V-shaped trough between the rows will float a graphite rod in the trough. One video on You Tube shows this done with a pencil "lead." (Those who have studied the diamagnetic properties of graphite rods from pencils have found that there is a great variation in magnetic properties from one type of pencil to another. If you try this experiment with a graphite rod from one type of pencil and it doesn't work, try one from a different type of pencil.) The same can be done with the trough formed between two rows of cylindrical neodymium magnets magnetized across their diameter. The Web page of Martin Simon of UCLA devoted to diamagnetic levitation shows a small piece of graphite floating near the top of the gap between two neodymium magnets taken out of a computer hard drive. Researchers have studied several other arrangements of magnets and floating graphite for use in seismometers to detect vibrations from earthquakes. And engineers from the Stanford Research Institute have patented the possible use of diamagnetic repulsion of magnets for magnetic bearings.

Repulsive forces between magnets and diamagnets have also been used to *stabilize* the levitation of small neodymium magnets. As discussed in Chapter 3, a magnet attracted upward by a fixed magnet above it can find a distance of separation at which the upward attractive magnetic force equals the downward gravitational force, but the delicate equilibrium of vertical forces is *unstable*. If the lower magnet is displaced slightly upward toward the upper magnet, the magnetic force of attraction will increase and exceed gravity, and the magnet will rise. If it is displaced slightly downward, the upward magnetic force will decrease and become less than gravity, and it will fall. Instead of restoring forces, there are *destabilizing forces*. However, if the lower magnet, positioned near that place of unstable equilibrium of forces, is placed between two slabs of diamagnetic material, the repulsion between the magnet and the diamagnetic slabs will resist displacement of the magnet upward or downward (will oppose the destabilizing forces), and stable levitation is achieved. Several scientific supply houses offer such devices, in which the upper attractive magnets are usually ferrite magnets (with a screw to adjust their height), the lower levitated magnet is typically a cube of neodymium magnet about 3 millimeters on a side, and the diamagnetic slabs of graphite are about 6 millimeters apart (Figure 15). In this case, the downward force of gravity on the lower magnet is balanced mostly by the attractive force of the upper magnet, and the graphite slabs provide only an additional small force to stabilize the magnet against vertical displacements. Such arrangements are therefore usually called diamagnetic stabilization rather than diamagnetic levitation. This device uses only easily available materials, requires no power input, can be easily passed around to students, and is an easy but impressive display of diamagnetism-aided levitation.

In this device, a strongly diamagnetic material, graphite, was necessary to stabilize the levitation of magnets. To successfully keep the magnets levitated, the diamagnetic repulsive forces acting on the magnets must be strong enough to counter the destabilizing forces resulting from displacements from the position of unstable equilibrium of forces. In situations where those displacement-induced destabilizing forces are smaller, even much weaker diamagnetic materials, like your fingertips, can produce enough repulsive force to levitate a magnet. Figure 16 shows such a case. Here the attractive upper magnet is a powerful (14 tesla) magnet 2.5 meters above the floating magnet. At such a large distance, the variation of upward attractive force with distance is rather slow, so vertical displacements result in only small departures from the equilibrium of gravitational and attractive magnetic forces, destabilizing forces are weak, and even the weak diamagnetism of fingertips can stabilize the magnet against such displacements. In such a case, a magnet can also be stably floated above almost any diamagnetic material, such as a book or a wooden desk, as can be seen on Martin Simon's UCLA website.

Figure 15. Diamagnetic stabilization of the levitation of a neodymium magnet by two graphite slabs. The large ferrite ring magnet at the top provides an upward attractive force to combat gravity, and diamagnetic repulsion from the graphite slabs stabilizes the levitation.

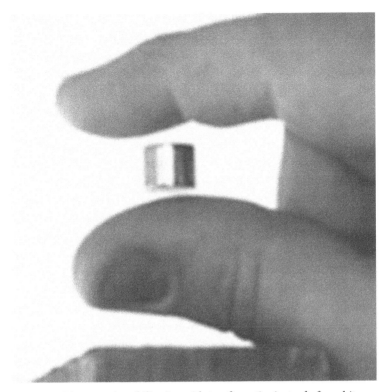

Figure 16. Diamagnetic stabilization with two fingertips instead of graphite. Here the upward attractive force on the neodymium magnet is provided by a powerful electromagnet high above the field of view. The upward attractive force on the magnet varies very slowly with position, so even the very weak diamagnetism of the two fingertips can provide repulsive forces sufficient to stabilize the levitation. The fingertips between which the magnet is levitating belong to Andre Geim, recipient of the Ig Nobel Prize for Physics in 2000 for frog levitation, and the actual Nobel Prize in Physics in 2010 for his research on graphene.

Froglev

Andre Geim and his colleagues at the University of Nijmegen became fascinated with the Japanese report of the Moses effect in an 8-tesla horizontal electromagnet. In their High Field Magnet Laboratory, they had vertical electromagnets of twice that strength and wondered what would happen if they simply tried to pour water down the central hole of their magnet. To their amazement, the water did not fall through the magnet to the floor, but remained stuck near the top of the magnet as long as the current to the magnet was flowing and the field was on. With a few minutes of calculations, based on the magnetic field and field gradient above their magnet and on the magnetic susceptibility of water, they realized that the upward diamagnetic forces produced by the electromagnet on the water near its top were sufficient to counteract gravity.

The next day they returned for further experiments and found that they could induce drops of water and many other objects, including wood, plastic, cheese, and pizza, to float near the top of their magnet. The word got around, and for most of the following week they found themselves demonstrating these wonders to a steady stream of visitors. Although most of the scientists at their university were vaguely aware of the diamagnetism of water and other common objects, they knew it was very weak, and they were amazed that it was strong enough to counteract gravity with the help of a very strong electromagnet. With a search of the scientific literature, the research group at Nijmegen soon learned, probably to their disappointment, that they were not the first to discover diamagnetic levitation. In fact, they learned that French scientists had demonstrated the diamagnetic levitation of water, wood, and plastic a few years earlier, a feat that had not generated much interest and was probably known to only a few. (So no Nobel Prize for discovering diamagnetic levitation. But in 2010, ten years after receiving his Ig Nobel, Geim received the Nobel Prize in Physics, along with a colleague, for later research on graphene, a single-layer form of graphite.)

Although they were not the first to discover the wonders of diamagnetic levitation, the Nijmegen group decided it was their duty to call the attention of both the scientific community and the general public to this magical phenomenon. They successfully levitated strawberries, small tomatoes, and even grasshoppers, but when they levi-

Figure 17. Diamagnetic levitation of a small living frog above the central hole of a high-field electromagnet.

tated a small frog and saw it kicking its legs and attempting to "swim" in the air above their electromagnet, they realized they had found the symbol and attention getter that they needed (Figure 17). It was a living thing and more appealing to most people than grasshoppers. Their photo of a levitating frog appeared in the April 1997 issues of *Physics World* and *New Scientist,* with respective headlines "Molecular Magnetism Takes Off" and "Frog Defies Gravity." The related press release led to articles in newspapers around the world and immediately drew wide public interest.

Among those interested was syndicated humor columnist Dave Barry, who responded in a column, "Get ready to dance naked in the streets, because scientists have done something that humanity has long dreamed about but that most of us thought would never happen. That's right: They have levitated a frog." Noting that the newspaper articles also said that the frog "showed no signs of distress after floating in the air inside a magnetic cylinder," Barry wrote, "Duh. Of course the frog 'showed no signs of distress': It's a frog. Frogs are not known for showing emotions; they are limited to essentially one facial expression, much like Jean-Claude Van Damme. What did these scientists expect the frog to do? Cry? Hop around on their computer keyboard and spell out the words, 'I am experiencing distress'?"

The news of the flying frog also stimulated a letter from someone in England claiming to be a representative of an unorthodox church, willing to buy Geim's magnet for one million pounds if it could be concealed under the floor so that, at one of his church services, he could "start to rise up from the ground and then (slowly and gently!) come back down again." He was sure that a demonstration of his ability to levitate would "get many more to join the church." "We have all this money," he wrote, "and we have the One True Word to save the world, but we have to do magic tricks to get the peoples to listen." The clever prank letter was apparently inspired by the Natural Law Party, formed by practitioners of Transcendental Meditation, who the writer noted "also do levitating," referring to their practice of "Yogic flying." (The Natural Law Party actually ran candidates for office several times in the United Kingdom and the United States, and in 1992 had George Harrison of the Beatles hold a fund-raising concert for them, but none were elected.)

A more serious person whose interest was aroused by news of the flying frog was Sir Michael Berry of the University of Bristol, England, who was among the physicists who had recently published a mathematical analysis of the stability of the spinning Levitron. The Levitron had defied Earnshaw's rule against magnetic levitation because it was spinning, and Berry also knew that Earnshaw's rule did not apply to diamagnetic materials. (This fact had been noted over a century earlier by William Thomson, commonly known as Lord Kelvin, but Kelvin thought that it would never become possible to produce magnetic fields strong enough to demonstrate diamagnetic levitation.) Berry soon contacted Geim and developed a mathematical analysis of the conditions for stable levitation of the frog, water, and similar diamagnetic materials. We noted earlier that the magnetic force on a diamagnetic material depended on the product of the magnetic field (teslas) and its gradient (teslas per meter). As Geim's group had estimated earlier, for the known density of water and its diamagnetic susceptibility, the upward magnetic force required to counter the downward force of gravity on the water was 1,400 tesla squared per meter. That condition was met near the top of Geim's electromagnet, at a local field of roughly 14 tesla and a field gradient at that position of about 100 tesla per meter—one tesla per centimeter. (In contrast, levitation of pyrolitic graphite is much easier and requires only about 60 tesla squared per meter.)

But balancing gravity was not sufficient for stable levitation. For both vertical and horizontal stability, it was also required to have the variation of the magnetic field with position provide restoring forces to resist both vertical and horizontal displacements from the equilibrium position. Geim measured the variation of the vertical magnetic field of his electromagnet with position along its axis, and plugging these data into the mathematical conditions derived by Berry showed that there should be just a narrow range of stability, of a bit less than an inch, near the top of the magnet. Geim's experimental observations fit well with Berry's theory. Berry had found mathematical similarities between the requirements for stable levitation of frogs and those he had derived earlier for Levitrons, and it was the paper by Berry and Geim entitled "Of Flying Frogs and Levitrons" that earned Berry his share of the 2000 Ig Nobel Prize for Physics.

In principle, you or I could also be levitated like the frog, since we have diamagnetic properties very similar to those of a frog. It would not require magnetic fields any higher than those in the electromagnet used by Geim, but that magnet had a central hole about 2 inches (5 cm) in diameter. To levitate even a very small person would require construction of a much larger and much more expensive electromagnet. Such a magnet has not yet been constructed. However, in his announcement of his acceptance of the Ig Nobel Prize, Sir Michael Berry wrote, "I have no reason to believe such levitation would be a harmful or painful experience, but of course nobody can be sure of this. Nevertheless, I would enthusiastically volunteer to be the first levitatee."

Dave Barry had expressed his doubts about the levitated frog's ability to communicate any "signs of distress" during and after levitation. Sir Michael is much more articulate than the frog, so if and when he is levitated, that concern of Dave Barry's will finally be answered.

Super-Levitation

Superdiamagnetism and Supercurrents

The diamagnetism of frogs is very weak, so levitation of the "flying frog" required an extremely strong magnetic field (and a strong field gradient). The diamagnetism of pyrolitic graphite can be fifty times stronger than that of frogs, allowing levitation of thin flakes of graphite about a millimeter (1/25 in.) above a set of neodymium magnets (Figure 14). But the diamagnetism of an ideal superconductor in low magnetic fields is about 20,000 times stronger than that of pyrolitic graphite! For an ideal superconductor, application of a low external magnetic field produces an opposing magnetic field *equal to* the applied field (a magnetic susceptibility of minus one), so that the net magnetic field inside the bulk of the superconductor is *zero*. The magnetic field penetrates only within a very thin surface layer, with the interior of the ideal superconductor completely shielded from the external applied magnetic field by persistent circulating surface currents in that layer. An ideal superconductor is *superdiamagnetic*. So with superconductors, magnetic levitation can be very easy indeed— once the material is cold enough to be superconducting. There's always a catch.

To achieve the remarkable phenomenon of superconductivity, you have to go down to super-cold temperatures. In discussing such temperatures, it is convenient to discard our familiar Fahrenheit scale and talk about temperature in degrees Kelvin (K). Kelvin degrees are the same size as Celsius (Centigrade) degrees, that is, nine-fifths of the more familiar Fahrenheit degrees, but we set the zero not at the freezing point of water, as we do for Celsius, but 273 degrees lower, at absolute zero, the lowest theoretically attainable temperature. In degrees Kelvin, water boils or condenses from gas at 373 K (100°C, 212°F). It freezes or melts at 273 K (0°C, 32°F). Nitrogen condenses at 77 K (minus 196°C) and freezes at 63 K. Helium, the most stubborn of gases, does not condense under ordinary pressures until 4.2 K.

Helium was first liquefied in 1908 by Dutch physicist Heike Kamerlingh Onnes, and soon he and his colleagues began measuring various properties of metals at super-low temperatures. Among other things, he measured *electrical resistance* as a function of temperature. Electrical current in amps, according to Ohm's law, is proportional to voltage, and the ratio of volts to amps is called the electrical resistance, measured in ohms. Onnes found that the resistance of metals decreased gradually as temperature was decreased, but he was amazed to find that below a temperature of about 4 K, the electrical resistance of a wire of mercury suddenly dropped to *zero*. Current continued to flow, but with zero voltage and zero resistance. The mercury was ohmless; it was a *super*conductor. Kamerlingh Onnes soon found that many other metals shared this amazing property, the electrical resistance suddenly dropping to zero below a *critical temperature*, although the critical temperatures for superconductivity were different for different metals (e.g., 7.2 K for lead, 4.16 K for mercury, and 3.72 K for tin). Of all the elements, the highest critical temperature for superconductivity is that for niobium, 8.7 K.

Kamerlingh Onnes was aware that the magnetic fields obtainable with electromagnets wound with copper wire were limited because of the heating produced by the electrical resistance of copper. The zero electrical resistance in superconductors suggested to him that an electromagnet wound with superconducting wire could produce record magnetic fields, but in this he was soon disappointed. He found that in the elemental metals he studied, superconductivity was destroyed with the application of even very modest magnetic fields, fields of only a few hundredths of a tesla (a few hundred gauss).

The metals he studied all lost their superconductivity not only when temperature exceeded their critical temperature, but also when the applied magnetic field exceeded their *critical field*. As the external applied magnetic field increased, the surface electrical currents that shielded the interior of the superconductor from magnetic fields also had to increase, but there was a limit to that superdiamagnetism. Once the applied magnetic field was above the critical field, the surface currents died, the magnetic field penetrated, and the material was no longer superconducting or superdiamagnetic. It converted to the "normal" metallic state, with normal nonzero electrical resistance.

If considering only increasing fields, superdiamagnetism can be explained with Faraday's and Lenz's laws of electromagnetic induction applied to a perfect conductor, that is, to a material with no electrical resistance. As discussed in Chapter 5, when copper or any normal conductor is exposed to an increasing magnetic field, eddy currents are induced that produce an *opposing* magnetic field (Lenz's law). But eddy currents in a normal conductor are limited by electrical resistance and die out quickly once the applied magnetic field is no longer varying with time. In a material with no electrical resistance, however, any induced currents can flow forever, and these surface "persistent currents" can continue forever to shield the interior of the material from magnetic fields. So superdiamagnetism in increasing external magnetic fields could be viewed simply as a result of perfect conductivity plus electromagnetic induction and Lenz's law.

In 1933, Walther Meissner and Robert Ochsenfeld showed that superdiamagnetism is more basic than that. Magnetic fields are excluded from the interior of a perfect superconductor not only in the case of a magnetic field increasing from zero, where it can be explained by Lenz's law (induced currents resist change), but in other situations as well. Suppose, for example, a magnetic field larger than the critical field is applied to an ideal superconductor, converting it to a normal metal and allowing magnetic field to penetrate the interior. If the applied field is then decreased below the critical field, the material converts back to superconductivity and is found to *expel* the existing magnetic field from its interior. Faraday and Lenz couldn't predict that! The same thing happens when an ideal superconductor is cooled from above its critical temperature in the presence of a small magnetic field. As soon as the material is cooled below the critical temperature, it becomes superconducting and *expels* the preexisting magnetic field. In these cases of decreasing field or decreasing temperature, instead of *resisting change* of the magnetic field within it, as Lenz's law would suggest, the superconductor *accomplishes* change by *expelling* the preexisting magnetic field. Apparently, whatever the history may be of the temperatures and magnetic fields applied to an ideal superconductor, magnetic fields are excluded from its interior. This is commonly called the *Meissner effect.* (Ochsenfeld lost out.)

From the Meissner effect, it is clear that a superconductor must have zero electrical resistance, since otherwise the surface currents excluding magnetic field from the interior would not be persistent. They would die out, and field would enter. So superconductivity (i.e., zero electrical resistance) can be inferred from the Meissner effect, but as we have seen, the reverse is not true. Perfect conductivity alone is not sufficient to produce the *expulsion* of magnetic field, the Meissner effect. Thus superdiamagnetism is more basic than superconductivity. (But the latter was discovered first, so the phenomenon is still called superconductivity.)

The most impressive demonstration of the persistence of supercurrents can be made with a simple ring of an elemental superconductor like lead. If you expose the ring to a perpendicular magnetic field at an elevated temperature and then cool it down to below its critical temperature, the Meissner effect will expel the magnetic field from the material of the ring itself, but not from the volume outside and inside the ring. If you then turn off the electromagnet, the field in the hole inside the ring cannot escape. The superconducting ring around the hole won't let it. There will be a net supercurrent circulating around the ring to maintain the magnetic field in the hole inside the ring. Researchers have used such an experiment to see just how small the resistance of an ideal superconductor really is. If there is any resistance at all, the net electrical current circulating around the ring will gradually decrease, and the magnetic field inside the ring will gradually decrease. But patient researchers have measured that magnetic field inside the lead ring for many months and have found no signs of any decrease, concluding that electrical resistance of an ideal superconductor is not just very small. It's zero! The current is ohmless, persistent, unchanging, a real *super*current. The superconducting current-carrying ring of lead is in effect a permanent magnet—as long as you keep it super-cold.

For the first half-century or so of research into superconductors, the materials studied most were nearly pure metallic elements like lead and tin. They behaved as we have described above, with perfect superconductivity and superdiamagnetism as long as the temperature was below the critical temperature and the applied magnetic field was below the critical field. Such superconductors are now called Type I superconductors, and it was with such materials that *superconducting levitation* was first demonstrated in 1945 by V. Arkadiev, a Russian scientist. He levitated a rod-shaped iron–nickel–aluminum permanent magnet above a concave dish-shaped bowl of lead cooled to 4.2 K by liquid helium. The lead bowl was a superdiamagnet and had persistent ohmless electrical currents—supercurrents—on the upper surface of the lead, producing a magnetic field opposing the field of the permanent magnet, and keeping the interior of the lead free from magnetic field. And the vertical repulsive force between the surface supercurrents and the permanent magnet levitated the magnet. The concave shape of the lead bowl provided stability against sideways displacements of the magnet because such displacements bring the magnet closer to the lead and produce increased surface currents and a sideways restoring force. (It is possible that it was the dish shape used by Arkadiev in this experiment that inspired Roy Harrigan to use a dish-shaped base magnet to stabilize the levitation of his spinning magnetic top that led to the development of the Levitron.)

The concave upper surface of the lead bowl provides a "magnetic well" that resists displacements of the magnet in any horizontal direction, much as the two triangular magnets in the base of the Revolution (Figures 4 and 6) resisted displacement in one horizontal direction. For the Levitron, a similar magnetic well was created in early designs by having the center of the base magnet magnetized in the opposite direction, and in some later designs by the use of a ring base magnet. For the floating aluminum pan (Figure 13) and the flying frog (Figure 17), the magnetic field from the electromagnets provided the magnetic well. In the Arkadiev experiment, supercurrents on the top surface of the lead bowl provide stabilizing restoring forces for four of the six degrees of freedom—three displacements plus rotation about a horizontal axis perpendicular to the rod magnet (such a rotation would bring one pole of the rod magnet closer to the surface). Rotations are completely unconstrained (neutral stability) about two axes—the vertical axis and the axis along the rod magnet.

It is worth stressing that in Arkadiev's experiment, the first demonstration of superconducting levitation, the magnet will *spontaneously lift* from the bowl when the system is cooled below the critical temperature of lead. A Type I superconductor like lead can exhibit a full-fledged Meissner effect—not just excluding magnetic field from its interior, but actually *expelling* the preexisting field. We will soon see that the superconductors of most interest today, Type II superconductors, rarely do that. They often don't even fully exclude the magnetic field. And that is often a very good thing.

In 1952, the second edition of David Shoenberg's influential book on superconductivity showed a permanent magnet levitating above a concave lead bowl cooled by liquid helium. He calls it the "floating magnet" experiment and wrote that it is "remarkable mainly for its aesthetic appeal," since he felt that the scientific explanation in terms of repulsion between a magnet and a superdiamagnet was fairly trivial. Although many practical applications of magnetic levitation have been developed in the intervening years, the "aesthetic appeal" of levitation still provides much of the motivation for its study today. Shortly after I arrived at the General Electric Research Laboratory in 1956, one of my new colleagues showed me an example of this classic levitation of a permanent magnet over a concave bowl of a Type I superconductor, and I found it fascinating. That was my first observation of the aesthetic appeal of fully contact-free magnetic levitation, and I remember it to this day.

Higher Fields

For the first fifty years of research into superconductivity, most materials studied were metallic elements like lead and tin, type I superconductors with low critical temperatures and low critical fields. But there were a few hints that this was not the whole story. In 1936, just three years after the discovery of the Meissner effect, a Russian scientist named Lev Schubnikow showed that several *alloys* of lead— lead mixed with other elements—lost their superdiamagnetism, that is, allowed some magnetic field to penetrate the material, at low fields, but still exhibited zero resistance to much higher fields. His research was cut short, however, because one year later, Schubnikow was declared an "enemy of the people" and executed in one of Stalin's many purges. Schubnikow's research results showing that it was possible to have supercurrents without superdiamagnetism (i.e., without full exclusion of magnetic fields) were prominently displayed in Shoenberg's book, but they remained a puzzling mystery until the 1960s.

It was in 1961, fifty years after the discovery of superconductivity by Kamerlingh Onnes and twenty-five years after the mysterious results of Schubnikow, that a group of researchers at Bell Laboratories reported that a compound of niobium and tin could carry high supercurrents even in the presence of very high magnetic fields. From then on, superconductors were no longer just materials of interest mostly to basic scientists, but materials of practical engineering value. The research group that I was working in at General Electric, and many other research groups around the world, immediately started studying what became called Type II superconductors, alloys and compounds that could carry supercurrents even after some magnetic field had penetrated. And soon many laboratories around the world, including ours, were able to realize Kamerlingh Onnes's original dream of constructing high-field electromagnets using superconducting windings. Today there are many thousands of superconducting electromagnets in use around the world, including laboratory research magnets (like the 16-tesla magnet with which Geim levitated that frog), hospital MRI magnets, and the miles of superconducting magnets arrayed in circular tunnels to guide high-energy beams around particle accelerators. With iron cores and ordinary copper windings, electromagnets had been limited to about 3 tesla. Soon electromagnets wound with superconducting niobium–titanium alloys were producing fields up to 9 tesla and electromagnets wound with niobium tin and related compounds were producing fields up to about 20 tesla.

The current world record holder for steady magnetic fields today is the 35-ton hybrid magnet (part copper, part superconducting) at the National High Field Magnet Laboratory of Florida State University in Tallahassee. It provides researchers with a magnetic field of 45 tesla (450,000 gauss, more than a million times the earth's field). This record-holding magnet consists of an insert of water-cooled copper conductors specifically designed to withstand huge magnetic forces, and a superconducting "outsert" wound with niobium titanium and niobium tin. The insert copper electromagnet is cooled by 4,000 gallons of water a minute and uses 35 megawatts of electrical power to produce 33.5 tesla. The outsert superconducting magnet adds 11.5 tesla to reach a total field of 45 tesla, but requires little power beyond that involved with keeping it cooled to liquid helium temperatures. After all, it's superconducting and has no electrical resistance.

Type I superconductors like lead and tin totally exclude magnetic fields less than their critical fields. Type II superconductors like niobium titanium and niobium tin and the lead alloys of Schubnikow are more compliant and use the "bend so you don't break" philosophy. They maintain some superconductivity up to high magnetic fields by allowing partial field penetration. Magnetic field penetrates the Type II superconductors in the form of linear *flux lines* that have tiny "normal" cores in which the local electron properties are like those in a normal conductor, surrounded by superconducting material carrying rings of supercurrent around the tiny normal cores. (In magnetism, the terms *flux* and *field* are closely related. *Magnetic field* commonly refers to the *intensity* of magnetic flux, i.e., the flux per unit area.)

Superconductivity is a quantum-mechanical phenomenon, and the total amount of magnetic flux carried in each flux line is quantized—that is, it has the same fixed and minimum value. (Our money is also quantized—into pennies.) As more and more magnetic field penetrates the material, it penetrates in the form of more and more quantized flux lines, which therefore get closer and closer together as the internal magnetic field increases. At low internal magnetic fields, the flux lines are far apart and the normal cores are only a small fraction of the total volume, but as the field increases, the proportion of normal material gradually increases, approaching 100% at a critical field that is characteristic for each material. So even Type II superconductors have critical fields beyond which they are no longer capable of superconducting, but their critical fields can be as much as several thousand times higher than the critical fields of Type I superconductors.

A Type II superconductor with partial field penetration is said to be in a "mixed state," since some of the material is locally superconducting but some of the material (the cores of the flux lines) is locally normal and nonsuperconducting. Will the bulk material with a mixture of superconducting and normal (nonsuperconducting) regions actually carry electrical current without resistance, actually be superconducting in the mixed state? That depends! Researchers in the 1960s, including me, learned that the ability of a Type II superconductor to carry supercurrent (i.e., current without resistance) in the mixed state depends on details of how the atoms of the material are arranged on the microscopic scale, that is, on its *microstructure*.

Paradoxically, the more imperfect the microstructure, the more supercurrent a material can often carry in the mixed state. When current is flowing in the mixed state, it exerts a transverse force on the flux lines. If the microstructure of the material is perfectly uniform on the microscopic scale, the flux lines can move easily in response to that transverse force, and the motion of flux lines produces electrical resistance in the form of local heating in the normal cores. Then there's little current flow without resistance, little supercurrent. But if the microstructure is not uniform but inhomogeneous, local irregularities like small particles of different chemical composition can block the easy motion of flux lines. Flux lines "pinned" on inhomogeneities can resist the transverse forces exerted by electrical currents and substantial current can be carried without resistance. The maximum current that can be carried in the mixed state before flux lines break away from pinning sites and create electrical resistance is called the *critical current*. Critical currents in the mixed state of Type II superconductors are increased by *flux pinning* by inhomogeneities, and materials engineers soon learned, partly by trial and error and partly by design, how to produce Type II superconductors that could carry large supercurrents at large magnetic fields.

The pinning of flux lines by chemical and physical inhomogeneities is thus the secret of the commercial success of Type II superconductors. Type I superconductors, remarkable though they are, can carry supercurrents only on their surfaces. But Type II superconductors with flux pinning can carry supercurrents throughout their interior, throughout their entire cross section. The ability to carry high supercurrents in the presence of high magnetic fields makes them ideal materials for the production of high-field electromagnets—and, as it turns out, for improved levitation.

The distribution of supercurrents and of flux lines in the interior of a Type II superconductor with flux pinning can be very complex, but a useful theoretical model to understand them was developed by Charles Bean, a close colleague of mine at GE. It is now called the Bean model, and it starts with one of Maxwell's equations, the Ampere-Maxwell equation. That equation relates gradients in internal magnetic fields, that is, changes in the number of flux lines per unit area, to internal currents. Adding the assumption that the current at each point within the material is always the maximum supercurrent that the material is capable of, it is possible to calculate internal field and current distributions for any complex history of applied external fields and currents. And, as we shall soon see, the Bean model can explain how magnets interacting with Type II superconductors with flux pinning can develop both attractive and repulsive forces between them and thereby achieve stable magnetic levitation in both repulsive and attractive modes.

Higher Temperatures

The ability of Type II superconductors such as niobium titanium and niobium tin to carry high critical currents in the presence of high magnetic fields finally led to important engineering applications of superconductivity, including products important to the general public in the form of electromagnets for MRI. But critical temperatures remained below about 20 K, and superconducting devices had to be cooled to or near the temperature of liquid helium, only a few degrees above absolute zero, to operate. The next revolution in superconductivity came with the discovery of materials with much higher critical temperatures. The breakthrough that excited both scientists and engineers, and even the *New York Times,* was the report in 1987 of superconductivity up to 90 K in a compound of four elements—yttrium, barium, copper, and oxygen—soon commonly abbreviated to YBCO. That's still super-cold by ordinary standards but well above 77 K, the boiling point of liquid nitrogen, a liquid much easier to work with, and much cheaper, than liquid helium.

Although I saw a magnet levitating above a superconducting lead bowl cooled with liquid helium over a half-century ago now, it was within a large thermally insulated container with only a limited transparent area through which I could peek to see superconducting levitation. After 1987, I could simply place a black disc of YBCO in a shallow glass dish, pour liquid nitrogen over it until the disc cooled to 77 K, pick up a neodymium magnet and place it over the YBCO disc, and see the magnet float stably about a half-inch above the disc. I could show this to students, and they could handle the magnet themselves, experiencing and exploring directly the magic of magnetic levitation. They could try different sizes and shapes of magnets, including a cylindrical magnet that could be spun easily about its axis, demonstrating the power of contact-free levitation in reducing friction. Their hands-on explorations with maglev above YBCO were a much more effective learning experience than a peek into a large thermally insulated container at liquid-helium temperature would have been. With the exception of the Levitron and the desk toys discussed in the next chapter, floating a magnet above YBCO is probably the demonstration of magnetic levitation that the most people have seen in person. (Lots of people have seen maglev trains, but the levitation is not clearly visible.) For those who haven't yet had the luck to see a magnet floating above YBCO, there are numerous videos available on You Tube. The repulsive force between the magnet and YBCO of course also allows you to float the YBCO above the magnet, but once out of the liquid nitrogen, the YBCO will soon warm above its critical temperature and lose its superconductivity.

Soon thousands of researchers around the world were studying YBCO and searching for similar materials that might have even higher critical temperatures. The maximum critical temperature so far found is 138 K, which is still super-cold (minus 135 Celsius, minus 211 Fahrenheit), but these newly discovered materials were quickly dubbed "high-temperature" superconductors, relegating all earlier superconductors to the diminished category of low-temperature superconductors. One compound of bismuth, strontium, calcium, copper, and oxygen, nicknamed BSCCO, was found easier to manufacture in wire form than YBCO, and has found some applications, but for demonstrations and applications of magnetic levitation, YBCO remains the most popular of the high-temperature superconductors.

Arkadiev's magnet levitating above lead in liquid helium required the concave shape of the lead bowl to provide lateral stability to the floating magnet. But the YBCO disc that I use with my students is flat, and nevertheless the levitated magnet is firmly stable against sideways displacements. Lead was a Type I superconductor, and surface supercurrents totally excluded the magnetic field of the magnet from the interior of the lead. However, YBCO is a Type II superconductor, and despite shielding supercurrents within the disc that repel and levitate the neodymium magnet, a portion of the field from the magnet penetrates the YBCO disc. Flux pinning within the disc obstructs the easy motion of flux lines within the disc, providing lateral stability to the magnet. Thanks to flux pinning, there's no need for a concave bowl with YBCO.

Researchers in Japan were among the first to develop methods of preparing YBCO with greatly enhanced flux pinning, which allowed stronger supercurrents and therefore stronger magnetic forces for levitation. By 1996, they had produced material that could levitate many pounds of weight. For a dramatic demonstration of the lifting power of their YBCO, they assembled the material into a disc about 2 feet in diameter, placed above it an assembly of neodymium magnets forming a disc of similar diameter, and published a photo of the smiling secretary of one of the researchers sitting on the magnet disc, which was levitated about an inch (2.5 cm) above the YBCO disc. It was a striking photo, but Japanese women tend to be rather small and light. Their next photo was more spectacular and showed the similar levitation of a sumo wrestler weighing over 310 pounds (140 kg). He was standing on a disc of magnets weighing about 130 pounds, so the total levitated weight was more than 440 pounds (200 kg), floating about an inch above the nitrogen-cooled YBCO. This impressive demonstration of superconducting levitation drew worldwide attention and is now featured in several books, including

Figure 18. Superconducting levitation of sumo wrestler Tosanoumi. He stands on a support plate of neodymium magnets that are repelled upward by an array of nitrogen-cooled inch-thick high-temperature superconductor (YBCO). The total levitated weight is 202 kilograms. Photo courtesy of Nihon Sumo Kyokai and ISTEC.

the present one (Figure 18). It also was featured on the levitation Web page of the Nijmegen high-field magnet laboratory, leading the Ig Nobel Prize committee, as noted in the previous chapter, to mistakenly credit Andre Geim with using magnets to levitate a sumo wrestler as well as a frog.

The spinning Levitron, the aluminum frying pan in Figure 13, the sumo wrestler, and Andre Geim's frog were all levitated by *repulsive* magnetic forces. As discussed in Chapter 3, levitation with *attractive* magnetic forces is usually vertically unstable. Thus imagine the surprise of researchers in the late 1980s when they found that a disc of YBCO could stably float *below* a neodymium magnet, a phenomenon that some have called "the suspension effect" (Figure 19a). Type II superconductors with adequate flux pinning can do wondrous things.

When you place a magnet above a disc of YBCO, the field from the magnet induces supercurrents in the disc that produce a field opposing the magnet field, consistent with the dictates of Lenz's law. The resulting repulsive force levitates the magnet. No surprise there. Push the magnet down a little closer, increasing the external field applied to the YBCO disc, and you will induce repelling supercurrents deeper into the disc and a greater repulsive force upward. But now move the magnet back up. Magnetic fields from a magnet decrease with distance of separation, so moving the magnet upward will decrease the magnet's field applied to the disc and will induce supercurrents *in the opposite direction,* currents that produce a field that *attracts* the magnet, again consistent with Lenz's law. Induced currents *resist change.* This is reminiscent of Faraday's original experiment in which currents were induced in a coil of wire both when a magnet was inserted in the coil ("when we meet") and when it was withdrawn ("when we part"). When Faraday inserted the magnet in the coil, the induced currents produced a field repelling the magnet, but when the magnet was withdrawn, currents were induced in the coil *in the opposite direction,* resisting change and attempting to attract the magnet back into the coil. The same thing happens in the YBCO, but the currents inside the YBCO are very different from the

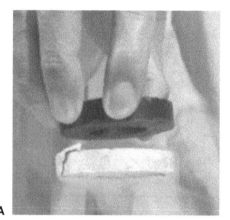

A

B

Figure 19. A disc of high-temperature superconductor (YBCO) floating (a) below, (b) alongside, and (c) above a neodymium magnet. The YBCO was cooled below its critical temperature with liquid nitrogen while in the field from the magnet, and the magnetic flux from the magnet was trapped inside the YBCO by flux pinning. The YBCO is wrapped in a thermal sheet to keep it cooled below its transition temperature.

C

currents in Faraday's copper coil, which rapidly decayed because of the electrical resistance of copper. Now the induced currents, both the initial repelling currents and the new attracting currents flowing in the opposite direction, are supercurrents—they both persist, although in different parts of the disc interior. The Bean model, applied to the case of first increasing and then decreasing field, can be used to determine the internal distribution of the two currents. When the force from the attracting currents outweighs the force from the repelling currents, the net attractive force can overcome gravity and the disc, still carrying both attracting and repelling supercurrents in different parts of its interior, will rise. And, as long as it remains cold, the YBCO disc, levitated by a net attractive force, will be stable.

How do we explain the vertical stability? We lean on Lenz's law and the Bean model. Consider the effect of vertical displacements from the position of equilibrium. If the disc were to rise closer to the magnet, it would experience an *increased* field from the magnet, and the increasing field would induce an increase in the *repelling* supercurrent, pushing it back down to the equilibrium spacing. On the other hand, if the disc were to move farther away from the magnet than the equilibrium spacing, it would experience a *decreased* field from the magnet, and the decreasing field would induce an increase in the *attracting* supercurrent. Vertical displacement in either direction produces a *restoring force*—the YBCO disc is vertically stable. The disc, containing both attracting and repelling supercurrents and capable of changing the balance between them in response to external field changes, is self-stabilizing!

An especially effective method to produce the "suspension effect" is to cool the YBCO down in the presence of a magnet held a short fixed distance above it, say on top of a nonmagnetic spacer. Cooling a superconductor down in the presence of a magnetic field is called field cooling. The field from the magnet freely penetrates the YBCO disc while the disc is above its critical temperature. When the disc cools below its critical temperature and becomes superconducting, it tries to expel the field, but if flux pinning is present, it cannot, and the disc traps much of the magnet's field within it. It now contains *trapped flux*. If you then remove the spacer holding the magnet and lift the magnet, the YBCO disc will lift with it. In this case the initial distribution of fields and supercurrents within the disc is somewhat different than in the earlier case, but the levitation or suspension will be vertically stable for the same reason. Displacements up or down will induce repelling or attracting supercurrents and resulting restoring forces. The suspension effect shown in Fig. 19a was actually achieved with field cooling. With the YBCO wrapped with a thermal insulator to keep it cold and remain superconducting longer, the magnet can be turned and the YBCO remains floating alongside (Figure 19b) or under (Figure 19c) the magnet.

A similar experiment has been done in which a sizable piece of thermally insulated YBCO was placed within a model train car and field-cooled above a permanent-magnet track. The track in this case consisted of three rows of magnets with an NSN pole pattern, that is, two rows of north poles surrounding a central row of south poles. Through flux pinning, field cooling traps within the YBCO, in the form of supercurrents, the complex field pattern emerging from the track. As a result of this internal field pattern, the YBCO is vertically stable *both above and below* the track, and stable against lateral displacements as well. The low friction and vertical and horizontal stability are sufficient to allow the train to travel smoothly along a curved track for a long distance after only a mild push, as can be seen on several videos available on You Tube. And the YBCO can be stably levitated not only above and below the track, but can even be stably levitated *sideways* (as in Figure 19b) if the track is turned vertical. In this orientation, it is the magnetic forces exerted from side to side *across* the track that directly oppose the force of gravity.

This is truly remarkable. Here the forces between the magnets in the track and the supercurrents in the YBCO stably maintain the relative position between the track and the YBCO, up, down, or sideways, with the force of gravity just a perturbation on the more powerful magnetic forces. This sheds light on an oversimplification often used about magnetic forces. If you consider the YBCO floating directly *above* the magnet track and focus only on the vertical forces, you can say that the downward force of gravity on the YBCO is balanced by a net *repulsive* force between the YBCO and the magnets. If you instead consider the YBCO floating directly *below* the magnet track, you can say that the force of gravity on the YBCO is balanced by a net *attractive* force between the YBCO and the magnets. But how do you describe it when the YBCO is floating *sideways* along a vertical track? Now it is clearer that the forces between the YBCO and the magnets are not directed along just one axis, and that forces in more than one direction are playing a role in producing stable levitation. Flux pinning in Type II superconductors and the patterns of supercurrents *resisting change* through Lenz's law can indeed produce remarkable results. Too bad we still have to keep the superconductor super-cold.

In Chapters 4 through 7, we have considered four ways in which Earnshaw's rule could be circumvented and contact-free magnetic levitation achieved:

1. Spin-stabilized levitation (the Levitron)
2. Eddy currents, induced via relative motion or AC fields
3. Diamagnetic materials (like frogs)
4. Superconductors, especially Type II superconductors with strong flux pinning

Each of these approaches can be useful in specialized applications. But each is too restrictive to enable full contact-free magnetic levitation in the most general case. For this, we need another concept that we introduce in the next chapter. We need feedback.

Feeding Back

Brief Recap

Up to this point, we have introduced several different types of magnetic forces. They include:

A. Attractive forces between magnets and iron
B. Attractive and repulsive forces between magnets and other magnets
C. Repulsive forces between magnets and conductors with eddy currents induced either by (1) relative motion or (2) pulsed or alternating currents (AC)
D. Repulsive forces between magnets and diamagnetic materials
E. Repulsive forces between magnets and Type I superconductors
F. Repulsive and attractive forces between magnets and Type II superconductors

In this list, "magnets" should be interpreted as including both permanent magnets and electromagnets, with or without iron cores. The term *iron* could be interpreted as including other temporary-magnet materials such as cobalt and nickel and various magnetic alloys, but they're used much less often than iron. (For completeness, I could also have included attractive forces between magnets and paramagnetic materials, but we have not discussed them because they are of very limited interest with regard to magnetic levitation.) The magnetic forces of most general application are types A,

B, and C. Type D forces are usually of interest only at very high magnetic fields, and types E and F only at very low temperatures. But in those limited cases, I hope that from reading Chapters 6 and 7 you will agree that they can sometimes be very interesting!

We have no more different kinds of magnetic forces to add to the list, but we need to introduce a concept that is very important to many practical cases of magnetic levitation. That concept is *feedback*.

Floating Globes

Figure 20 shows an example of what has become a fairly common "desk toy"—a floating globe. (When I was demonstrating simple forms of magnetic levitation to my granddaughter's first-grade class, one boy was less impressed than the other students. He informed me that he had a floating globe in his house.) Such globes usually contain magnets at their north and south poles, and above the globe, at the end of a long curved arm attached to the base, is a bit of metal. Within that metal is a magnet that produces an attractive upward force on the magnet in the globe's north pole, equal and opposite to the downward force of gravity on the globe, and that equilibrium of forces allows the globe to float in the air contact-free. Sounds simple enough, but isn't levitation by attraction vertically unstable? Yes, it is ordinarily, but the seemingly simple desk toy provides stability by a method not discussed in the previous chapters, a method used in most industrial uses of magnetic levitation. It provides vertical stability through the magic of *feedback*.

The secret of maintaining stable levitation of the globe is that the magnet attracting it upward includes an electromagnet, and the current to that electromagnet is being varied many times a second. If the globe starts to fall a bit *below* the equilibrium position, the current to the electromagnet is increased, yielding a net restoring force (magnetic force minus gravity) *upward*. If instead the globe starts to lift a bit *above* the equilibrium position, the current to the

Figure 20. Floating globe. Levitation by an upward attractive magnetic force stabilized by sensor and feedback circuit.

electromagnet is decreased, producing a net restoring force (gravity minus magnetic force) *downward*. Thus departures from the equilibrium position are opposed by *restoring forces*. Determining whether the globe is below or above the precise equilibrium position is the job of the *position sensor*—a device that measures the instantaneous vertical position of the globe. That information is delivered from the position sensor to the electromagnet through an electrical "feedback circuit" designed so that the current to the electromagnet is increased or decreased precisely as needed to keep the globe levitated in the equilibrium position where the magnetic force and gravity are equal and opposite. Systems like this in which sensing of the "output" (in this case, the vertical position of the globe) is fed back to the "input" (the current to the electromagnet) are called *servomechanisms* or, more simply, *servos*. This servo or feedback process is going on many times a second to keep the globe floating in the desk toy, a toy that is obviously not quite as simple as it looks.

Feedback control is ubiquitous in today's technology. Perhaps the simplest example is the thermostat that controls the temperature in your house. Here the input is the heat from your furnace, the desired output is the household temperature, and the sensor is the thermometer in your thermostat, which turns the furnace on and off if the temperature is below or above the desired temperature you have set. Another familiar example is the cruise control in your car. The input here is gas to the engine, controlled by the throttle, and the desired output is the speed of the car, measured by a rotation sensor attached to the output shaft of the transmission. (The rotation or speed sensor is commonly based on a small magnet mounted on the shaft that induces an eddy-current pulse in a stationary coil each time it moves past the coil. The frequency of induced current pulses measures the rotation speed of the shaft, hence the speed of your car.) The feedback control system compares the measured car speed to the desired set speed and accelerates or decelerates the car accordingly.

Feedback control is important for much of magnetic levitation because Reverend Earnshaw tells us that at least one of the six possible degrees of freedom of the levitated object (three directions of motion plus three rotations) will be unstable, and therefore needs to be actively controlled. We escaped his rule in situations where his rule did not apply—spin stabilization, eddy currents, and with dia-magnets or superconductors—but in most cases, we need feedback control of at least one of the degrees of freedom for full stability. In the case of the floating globe attracted upward, it is the vertical dis-placement that is unstable without feedback control.

Today there are many available maglev items similar to the float-ing globe in Figure 20, all operating by feedback to provide vertical stability. There are numerous versions of the planet earth, including a variety of colors and even one "zoo globe" featuring the animals na-tive to each region of the earth. Some also rotate, like the earth itself, with the help of rotating magnetic fields. A set named "space mission" features the earth, the moon, and Mars, while a "zero-G sports" col-lection includes a floating soccer ball, baseball, and basketball. Many nonspherical floating objects are also available, including vintage cars and buses, airplanes (747s, 777s, or DC-3s), the space shuttle, digital clocks, and picture frames. Whatever you may want to float, it's prob-ably available somewhere.

It is safe to assume that each of the floating objects, be it the earth, Mars, a baseball, or a miniature 747, contains a magnet. In most of the systems, the position sensor is a *Hall sensor* (discussed in Chapter 4), located above or below the floating object, which detects the magnetic field from the magnet in the object. As the object falls or rises away from its equilibrium position, the resulting decrease or increase in the field at the Hall sensor (magnetic fields vary with dis-tance of separation!) changes the voltage across it, and the feedback circuit translates that into an increase or a decrease in the current to the electromagnet.

The floating globe on my own desk at MIT uses a different type of sensor to detect the vertical position of the globe—an *optical sen-sor* in the form of an invisible infrared laser beam, akin to the infra-red beam that operates your television from your remote. (Infrared light has a longer wavelength and lower frequency than visible light.) Instead of a curved arm from the base supporting the attracting electromagnet, the electromagnet is supported by two vertical rods, one on each side of the floating globe. Each of the vertical rods has a short horizontal extension near the top of the floating globe, and the laser beam goes from one side to the other. If the globe falls a bit *below* the equilibrium position, less of the laser beam is blocked, and the increased intensity of the beam at the receiving end is converted, via a *photodetector* (a device that converts light intensity into electric current) and the feedback circuit, to an *increase* in current to the electromagnet. If the globe rises *above* the equilibrium position, more of the beam is blocked, and the decreased intensity of the received laser beam is converted into a *decrease* in current to the electromagnet. Although the position sensor in this device is an optical (light-based) sensor instead of a Hall (field-based) sensor, it is still the feedback circuit that provides vertical stability to the magnetically levitated globe by controlling the current to the overhead attracting electromagnet. Electromagnets, and their ability to change magnetic fields through changing electric currents, are basic to feedback-controlled magnetic levitation.

Hall sensors and optical sensors are among the most commonly used position sensors in maglev systems. Others include capacitive sensors and inductive sensors, dependent on particular properties of AC electric circuits, and eddy-current sensors, based on currents induced in conductors by AC fields. Such properties often depend not on the absolute position of an object in space, but specifically on the separation or gap between one object and another, where they are often referred to as *gap sensors* rather than position sensors. More about sensors will be presented in this and the following chapters.

An engineer named Guy Marsden offers for sale on the Internet a simple levitation kit that employs a Hall sensor. Although it has a visible attracting electromagnet above the levitated object and a small but visible Hall sensor at the bottom of the electromagnet, he writes that he decided against an optical sensor because of "the visual 'give away' of the light beam's components. We're going for magic here—right?" With a visible sender and receiver of a laser beam surrounding the levitated object, he felt that the effect would be not quite as magical, and it is the magical effect that we are going for. Marsden notes that the phenomenon of levitation "inspires images of stage magicians and lovely ladies." For readers of Harry Potter, levitation inspires thoughts of magic wands and incantations, but Marsden quotes famed science fiction writer Arthur C. Clarke: "Any sufficiently advanced technology is indistinguishable from magic." In this case, it is the magic of magnetic levitation.

Like the Levitron, the floating globes are "only a toy," but the aesthetic appeal and magical visual effect of contact-free floating objects have made both types of toy very popular. Recently a new type of maglev toy has appeared on the market, and I was one of the first to buy one. (I was already thinking of writing this book.) It was developed by a French company (Simerlab), manufactured in China, and marketed in the United States by Fascinations, the company that markets Levitrons. That probably explains why it is named Levitron AG (for antigravity) even though the underlying physics is totally different. The Levitron AG (Figure 21) features a floating globe containing a magnet at the south pole, but there is no magnet in the north pole and no attracting electromagnet above the globe. There is only a curved base underneath the globe that contains all the electromagnetic components necessary to provide stable levitation, this time with an upward net *repulsive* magnetic force opposing gravity. The lack of an overhead electromagnet and supporting arm in the Levitron AG increases the magical effect and aesthetic appeal of the floating globe, which is why they can get away with charging more for the Levitron AG than for most floating globes.

Like the earlier Levitron, the Levitron AG has led to a patent fight. Mike and Karen Sherlock and their company Levitation Arts had purchased rights to a 1992 patent by Lorne Whitehead of the University of British Columbia, Canada, and they claim that the Levitron AG violates the Whitehead patent. Since the Levitron AG is mar-

Figure 21. Levitron Ion AG (antigravity) Globe. The Ion version, as indicated by the arrow, rotates as well as levitates. The globe has a permanent magnet at its south pole, but all of the other magnets, sensors, and feedback circuits are within the base below.

keted by Bill Hones of Fascinations, this once again pit the Sherlocks against Hones in the law courts, and led to an early result that favored the Sherlocks. However, several issues remain. Comparison of the 1992 Whitehead patent and the 2007 Simerlab patent application reveals similarities in basic principles but differences in detail, and a judge may ultimately have to decide the relative importance of the similarities and the differences.

How does the Levitron AG work? For the floating globe of Figure 20, levitated by attraction, the sensor and feedback circuit only has to stabilize the vertical position. The Levitron AG, Figure 21, levitated by repulsion, is vertically stable but horizontally unstable, which requires a more complex sensor and feedback system. I didn't want to disassemble mine—it cost almost a hundred dollars—but I did examine it externally with a compass and a magnetic viewing strip, and I suspect that the internal structure is something like that shown in the various diagrams in the Simerlab patent application. The floating globe contains a strong permanent magnet at its south pole, while the base contains a ring magnet that produces an upward repulsive force at the separation between globe and base of almost an inch. In the first figure of the patent application, within the ring magnet is an array of six iron-core electromagnets and three Hall sensors. (A colleague who has disassembled a recent Levitron AG tells me it contains four electromagnets and two sensors.) But in the Levitron AG, except for the magnet in the south pole of the floating globe, all the components necessary to produce stable levitation of the globe are present in the base—permanent magnets, a set of sensors to detect departures from the equilibrium position, and electromagnets to deliver, in response to sensor signals delivered via feedback circuits, changes in magnetic fields to produce stabilizing restoring forces to the floating globe.

The 1992 Whitehead patent for a "levitation system with permanent magnets and coils" shows different physical arrangements of magnets, coils, and sensors in the base, but of course it is based on the same physics and a similar type of feedback control circuit, as are other earlier and later related patents. An interesting part of the Whitehead patent specifically states that "the force of gravity is not a key part of the operation of the present invention . . . the device of the present invention may be turned to any angle relative to gravity and the levitated element will stay in a stable position relative to the levitating device." A Levitation Arts video on You Tube indeed shows a Whitehead device in which a magnet is first levitating above the base, but when the base is rotated sideways and then upside down, the magnet remains stably levitating alongside and then underneath the base. This is reminiscent of the YBCO superconductor described in the previous chapter (Figure 19) that was able to levitate above, below, or alongside a magnet. (And the YBCO can do that without the benefit of sensors and feedback circuits!) I have found no claim about gravity being "not a key part" in the Simerlab patent application, although they do have an online video showing a propeller levitating and turning in midair alongside a wall-mounted base (i.e., a sideways levitation). That encouraged me to try to tip my Levitron AG, but it doesn't seem to work at much of an angle.

The Levitation Arts video also shows a large table over which objects not only can be levitated but also can be moved horizontally over distances of several feet with a controlling joystick. For example, it shows a levitating ball moving a complex path through an arrangement of many fixed hoops. Probably the visible motion of the levitated ball was produced by the unseen motion, under the table, of the levitator and all its magnets, sensors, and feedback control circuits. Unfortunately, at time of writing no products are yet on the market from Levitation Arts.

Fascinations now also offers on its website the Levitron Revolution (combining the names of two prior and popular toys), a platform from Simerlab on which you can levitate any object you want—as long as it is no heavier than 12 ounces. A generic levitation platform is also produced by a Dutch company called Crealev. Their module CLM-1 consists of a circular floating part about 5 inches in diameter, containing a hexagonal ring of permanent magnets, and a slightly larger flat base containing all the necessary components for magnetic levitation—permanent magnets (the magnetic viewing strip reveals four in a square array), electromagnets, sensors, and feedback circuits. If you pass a piece of paper under the floating platform, as you can do with the floating globe of the Levitron AG with no effect, the CLM-1 platform falls. Crealev clearly uses a vertical light beam between base and platform as part of its sensor system to measure and control levitation height. Unloaded, the CLM-1 platform levitates above the base more than an inch. Holding the specified maximum load of 1 kilogram (about 2.2 lb), the levitation height is a bit over half an inch.

Figure 22. "Floating Sculpture, 2008–2009" by Jane Philbrick. Twelve balls rest on twelve levitated platforms containing permanent magnets, above twelve bases containing most of the magnets, sensors, and circuits. Courtesy of Jane Philbrick and Skissernas Museum, Lund, Sweden. Photo by Christina Knutsson.

That's a much higher maximum load than the 12 ounces for the Levitron AG, but the CLM-1 is also much more expensive.

Jane Philbrick, a visiting artist at MIT's Center for Advanced Visual Studies, consulted with me in 2008 about her art projects and became fascinated with magnetic levitation. She used twelve Crealev maglev modules to float an array of twelve red balls in her striking "Floating Sculpture, 2008–09" that was a prominent part of her solo exhibition at the Skissernas Museum in Lund, Sweden in the fall of 2009 (Figure 22). Jane juggles many creative art projects at a time and, like the Crealev maglev modules, is capable of "keeping many balls in the air."

Floatin' in the Wind

As noted in Chapter 1, it was in December 1903 that the Wright brothers flew the first successful heavier-than-air machine. But as Wilbur Wright himself has written, "it all would never have happened if we had not developed our own wind tunnel and derived our own correct aerodynamic data." The basic idea of the wind tunnel, determining forces on vehicles moving through still air by using still models in moving air, was invented several decades earlier, but the Wrights were among the first to collect reliable data from wind-tunnel tests. It was in the winter of 1901–1902 that they built their simple wind tunnel, a pine box 6 feet long and 16 inches across, and used a gas-powered fan to blow air through the box at speeds up to 20 miles per hour. (The Wrights referred to the two ends of their wind tunnel as the "goesinta" and the "goesouta.") In that box they mounted, one after the other, over 200 models of different wing designs, including monoplanes, biplanes, and triplanes, models between 3 and 9 inches long. Through the mount support, they systematically recorded the lift and drag forces associated with each design, and the data they gathered helped them choose the wing design for their "Flyer" that two years later would make history at Kitty Hawk, North Carolina. But according to Wilbur Wright, without their prior wind-tunnel work, the Flyer would never have gotten off the ground.

The success of the Wrights led to the construction of increasingly elaborate wind tunnels, with increasingly sophisticated measuring instruments, to match the needs of the growing aircraft industry for testing designs of new airplanes. MIT used wind tunnels for testing and for instruction in aeronautical engineering as early as 1896, and in 1938 it completed the construction of an advanced wind tunnel called the Wright Brothers Wind Tunnel, just in time for extensive testing of military aircraft designs during World War II. Gene Covert was in the U.S. Navy during and after the war, but he joined MIT in 1952. And later in that decade, he and his colleagues developed for the Wright Brothers Wind Tunnel the world's first practical wind-tunnel magnetic levitation system (although they called it a Magnetic Suspension and Balance System).

Wind-tunnel testing of models of course requires some sort of support structure to hold the model in place as the wind rushes past it. But researchers had learned that mechanical support structures can interfere substantially with the air flow, in some cases even producing a drag force on the support wires that was more than the drag force on the model. The solution? Remove the support wires and support the plane model with magnetic fields! The system that Covert and colleagues developed was identical in principle to the gadget that floats a globe on my desk. They used electromagnets to provide magnetic forces to combat gravity, sensors that measured the position of the floating object, and a feedback circuit to provide stability by using the sensor information to control the currents to the electromagnets. The subscale models they levitated were not much larger and much heavier than my globe, but for their wind-tunnel applications they needed a larger separation between the electromagnets and the floating objects than provided in my desk toy, so they used larger electromagnets and higher magnetic fields. And their systems used both electromagnetic (inductive) position sensors and optical sensors. The electrical properties of an AC electromagnet (current-voltage relations) are more complex than current-voltage relations of conductors in DC (direct current) circuits. Part of the ratio of alternating voltage to alternating current depends on frequency (Faraday's law—since the rate of change of magnetic fields increases with increasing frequency, the voltage does as well) and is called *inductance* rather than resistance. The inductance of an electromagnet is changed when there are ferromagnetic materials like iron in the range of their magnetic fields, and the closer the iron is to the electromagnet, the greater is that change. They placed iron rods within their models, and the change in electrical properties (inductance) of the electromagnets thereby became a measure of the position of the model, providing the feedback circuit the necessary information to stably float the object. The currents to the electromagnets required to hold the model in place in the wind also directly provided a measure of the various forces and torques (rotational forces) produced on the model by the wind. In 2005, Gene Covert was awarded the prestigious Guggenheim Medal for major contributions to aeronautics, including, among other things, development of a practical magnetic suspension system for wind tunnels.

The feedback circuit that floats the globe on my desk only has to provide stability in one direction—vertical. Ideally, the maglev systems for wind tunnels should control all six degrees of freedom of the floating object. Furthermore, from electromagnet currents required to hold the model in place in the wind, they should measure the forces associated with each of the three directions—lift, drag, and sideways forces—plus the torques associated with each of the three directions of rotation, which have their own names. Rotation around the axis of the levitated object is called *roll,* rotation around the vertical axis is called *yaw,* and rotation around the transverse axis (nose up or nose down) is called *pitch.* The original MIT electromagnetic sensing system could handle five of the six degrees of freedom, but couldn't handle roll, since it was based on an iron rod in the model, and rotations of the rod along its axis(i.e., roll) would not change its effect on the electromagnets. The MIT group later used metallic shielding of parts of the model and other approaches so that roll of the model would change the magnetic interaction between the iron rod and the electromagnets.

A maglev system developed at MIT was sent to NASA's Langley Research Center in Virginia and was used to test, among other things, models of the space shuttle (Figure 23). NASA eventually decided that the advantages of maglev support systems were outweighed by their increased cost and complexity, and today none of NASA's forty-seven wind tunnels use maglev systems. However, several maglev systems (still commonly called Magnetic Suspension and Balance Systems) are in use in wind tunnels around the world and are of course more advanced than the early system developed at MIT. Most have more

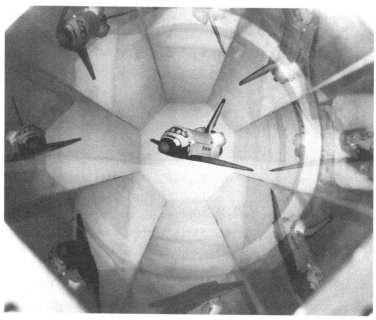

Figure 23. Small model of space shuttle floating in magnetic levitation system in wind tunnel at NASA Langley, with its image reflected in surrounding walls.

complex arrays of electromagnets, and most use neodymium permanent magnets in the models rather than iron rods. Several use optical
sensors in the form of vertical and horizontal light beams as the primary position sensors. One of the largest and most advanced systems
is at the Japan Aerospace Exploration Agency, which uses an array of
ten electromagnets to provide measurement and control of all six degrees of freedom. Behind and in front of the levitated objects are two
square electromagnets with central holes about 2 feet on a side to
allow free flow of air. Those magnets control the model's position
along the wind path and provide the primary measurement of drag
forces. On each side of the model are four more electromagnets, which
in combination provide and measure lift forces, sideways forces, and
the roll, yaw, and pitch torques. Researchers also use special cameras
and a fine dispersion of oil droplets in the air stream to measure the
air velocity as a function of position around the levitated objects, including simple shapes such as spheres and cylinders.

Magnetic levitation with feedback control has been used in wind
tunnels for more than a half-century, but there are many much newer
projects employing maglev with feedback. In the next sections we
sample a few.

Snowball in Hell

The Levitated Dipole Experiment (LDX) is a joint project of MIT
and Columbia University, located in the Plasma Science and Fusion
Center at MIT. There Darren Garnier of Columbia and his LDX colleagues stably levitate both very small and very large objects. On the
small end is their Levitated Cheerio Experiment (LCX) in which they
use an infrared laser beam and feedback control to levitate a small
block of wood about the size of a Cheerio containing a small neodymium magnet. This demonstration is designed simply to introduce to visitors the basic concept of sensor and feedback control to
achieve vertically stable magnetic levitation in attraction. After that,
visitors are ready to learn about the *much* larger object the plasma

researchers float stably—a superconducting electromagnet ring that, with its container, is about the size and shape of a large truck tire and weighs more than half a ton (Figure 24).

The LDX is designed to study the behavior of plasmas (ionized gases), usually of hydrogen isotopes, in the simple field of a magnetic dipole—a dipole in the form of a supercurrent circulating around a levitated ring wound with niobium–tin superconductor. The plasma, with temperatures of several million degrees, travels around the ther-

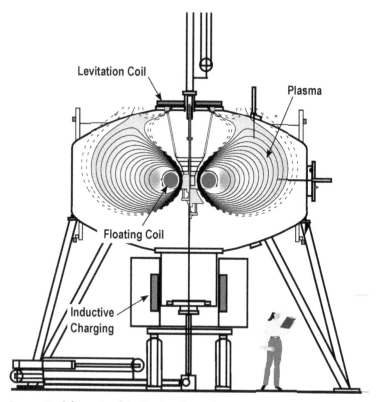

Figure 24. Schematic of the Levitated Dipole Experiment (LDX) at MIT, in which a 1,200-pound doughnut-shaped container holding a superconducting current-carrying ring is stably levitated for plasma studies.

mally insulated container surrounding the levitated superconducting ring. The ring, in its thermally insulated container called a cryo-
stat, is cooled to super-cold liquid-helium temperatures, just a few
degrees above absolute zero, but is surrounded by super-hot plasma,
so the researchers call it the proverbial "snowball in hell." During the
plasma experiment, the 550 kilograms (1,200 lb) ring is magnetically
levitated rather than mechanically supported because mechanical
support structures would disturb the plasma flow, much as mechanical support structures of an aircraft model in a wind tunnel can disturb air flow.

(Although our major interest here is the levitation, we should
note that high magnetic fields in various geometries are of general
importance to scientists studying fusion, the reaction in which two
hydrogen nuclei merge to form a helium nucleus, releasing lots of
energy. As a result of Fact 9, the magnetic force on moving charged
particles mentioned in Chapter 2, magnetic fields tend to contain
fast-moving ionized atoms within limited volumes, thereby increasing the likelihood of fusion reactions.)

Before the niobium–tin ring is lifted to where it will be levitated, it rests above another electromagnet, a "charging coil" wound
with superconducting niobium–titanium, located at the bottom of
the huge vacuum chamber within which the plasma experiments
will be done. While the niobium–tin ring is still above its critical
temperature, the charging coil is energized and bathes the niobium–
tin ring in a large magnetic field. After the niobium–tin is cooled in
that field and becomes superconducting (field cooling), the field
from the charging coil is turned off, and a huge persistent supercurrent is induced in the niobium–tin ring to maintain the field within
the ring (thank you, Faraday, Henry, and Lenz). The physics is essentially the same as the physics we discussed in the previous chapter about induced supercurrents in a lead ring, although because
niobium–tin is a Type II superconductor, some magnetic field also
penetrates the windings of the ring. And because the superconductor has strong flux pinning, it can carry very large currents indeed.

An induced supercurrent of over a million amperes flows around the niobium–tin ring, creating a maximum field within the ring of about 3.5 tesla.

The supercurrent-carrying niobium–tin ring is then mechanically raised to just below where it will be levitated for the plasma experiments, and an overhead "levitation coil" is turned on to levitate the niobium–tin ring with an upward attractive force. The original design called for a levitation coil wound with BSCCO, a high-temperature superconductor, but problems with that coil led to it being replaced with a water-cooled copper-wound electromagnet that consumes 150 kW of electrical power. This levitation coil produces at its center a field a little more than half a tesla and is about 1.5 meters (almost 5 feet) above the floating ring. Since this field falls off very rapidly with distance (see Chapter 3), in the vicinity of the floating ring it is only a small fraction of a tesla. Nevertheless, that modest magnetic field produces on the ring an upward vertical force enough to balance its weight, that is, about 550 kilograms (1,200 lb).

There are several different ways to think of that levitating force. You can think of the superconducting ring as a permanent magnet magnetized in the vertical direction, and the force resulting from the attraction between its upper pole and the lower pole of the levitation coil above it. From that viewpoint, the levitating force is produced by the upward gradient of the vertical component of the levitating field (in the vicinity of the ring) acting on the magnetic strength of the ring, called its magnetic moment. Although that field gradient (in teslas per meter) is much smaller than in most of the other maglev systems discussed in this book, the ring is carrying over a million amperes of supercurrent, so the magnetic moment of the ring is much larger than that of the levitated magnets in the other systems. And the product of the vertical field gradient and the magnetic moment of the ring yields a large levitation force. But if you prefer, you can instead think of the ring as a current-carrying wire and apply Fact 10 from Chapter 2: "A current-carrying wire in a perpendicular magnetic field experiences a force in a direction perpendicular to both the wire and field." Since here the current is flowing in a horizontal circle, it is the horizontal radial component of the levitation field that produces a force in the vertical direction, that is, in the direction perpendicular to both the current and that component of the field. Now the product of a small radial magnetic field and a huge circular current yields a large levitation force. With either way of thinking about it (or perhaps even not thinking about it), the magnetic interaction between the levitation coil and the ring results in a floating ring—a levitated dipole.

In the LDX, eight laser beams monitor the position of the floating ring and, through a complex feedback circuit, control the current in the levitation coil to keep the levitation vertically stable. Once stable levitation of the ring is achieved, the hydrogen isotopes can be let in to the vacuum chamber and heated to millions of degrees for the desired plasma experiments. All of this is made possible by something truly amazing—a floating 1,200-pound "snowball in hell" carrying a persistent circular supercurrent of over a million amperes. That's pretty cool in more ways than one.

Microbots and Pigs

Let's now travel from MIT to the University of Waterloo, Canada, and visit their Maglev Microrobotics Laboratory. As implied by the "micro" in their name, this group levitates much lighter weights than the LDX group, more in line with its levitated Cheerio than with its huge and heavy levitated superdipole. The Waterloo research team led by Behrad Khamesee uses a set of electromagnets, three laser beams, and sensors and feedback to levitate "flying microrobots" less than a gram (1/28 of an ounce) in weight. Each of their microrobots, or microbots, consists of two tiny neodymium magnets and an attached gripper that can be activated with a laser beam. Their electromagnets not only allow them to levitate their magnetic robots, but also to control their motion in space with considerable precision. Over a working volume about an inch on a side, they can control the position of a microbot to an accuracy of 20 microns (twenty millionths of a meter). The specialty of the miniature robots is micromanipulation. By controlling currents to the electromagnets, a robot can be moved to a precise location, the micro-gripper opened by the heat of a laser beam to pick up a very small object, the robot moved to another precise location, and the gripper opened again to deposit the tiny object in the new location. Without levitation, friction or adhesion would in general make it difficult to move with the precision these microbots are capable of.

In the preceding paragraph, I introduced the term *micron,* a unit commonly used by scientists and engineers when describing microscopic distances and dimensions. In future chapters we'll be encountering microns again, and even much smaller dimensions, so I think I should return briefly to the topic of units, which I discussed in Chapter 1. Until now, I have usually offered translations between inches and centimeters (0.39 in.) or millimeters (0.039 in.). But when I'm referring to dimensions much, much smaller than that, I'm not sure that translating them into inches will be of much help. This is, after all, scientific territory, so here I'll stick to metric units. The term *micron* is a shortened version of the full term *micrometer,* or millionth of a meter, and is abbreviated bilingually as μm. A millimeter, a thousandth of a meter, is already pretty small, about the length of the tail on the commas in this sentence. And a micron is a thousand times smaller than a millimeter. Our eyes can see things a millimeter in size (although I need my reading glasses), but we need a good microscope to see things that are only a micron in size. Think small.

Martin Simon of UCLA, an expert on Levitron science and on diamagnetic stabilization and levitation, also uses sensors and feedback to levitate magnetic objects, objects much bigger than the microbots at the University of Waterloo and much smaller than the superdipole at MIT. But he has levitated them inside a living pig! (see Figure 25).

Simon's colleague in this project, seen standing in the foreground in Figure 25, is Dr. Yoav Mintz of Hadassah University Hospital of

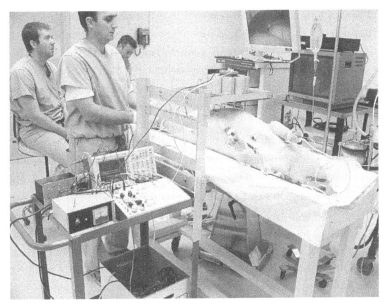

Figure 25. Maglev operation within the abdominal cavity of live pig at University of California San Diego. A magnet platform inside the pig is controlled by an array of electromagnets above the operating table and a Hall sensor below the pig.

Jerusalem. Mintz is optimistic about the future of magnetic levitation in surgery, particularly for performing diagnosis and surgical procedures inside the abdominal cavity. While visiting San Diego, he and physicians of the university hospital utilized a maglev system designed by Simon to demonstrate the feasibility of such operations by inserting a magnet platform into the abdominal cavity of a living pig, the cavity inflated with carbon dioxide to provide a large working volume. On a metal frame above the operating table was an array of electromagnets to levitate and move the magnet platform inside the pig in a controlled fashion, guided by feedback signals from Hall sensors underneath the pig. (With the levitated magnet platform inside the pig, they could not use external laser beams as position sensors.)

With a small camera also inserted into the pig's abdominal cavity, they directly observed their ability to move the magnet platform with the external electromagnets in a controllable fashion while it was inside the pig. In this preliminary experiment, the platform was inserted through an incision, but the eventual goal is laparoscopic or minimally invasive therapy, and a smaller magnet platform would be inserted through the mouth. The pig was sacrificed afterward, but Mintz and his medical colleagues considered the experiment a success. Simon has also designed and constructed a related maglev system in which the levitating magnets are not above the magnet platform but about a foot to each side of it. Here the dominant magnetic field is not directed vertically, but horizontally, and the upward magnetic force comes from the gradient or change in this horizontal field with vertical position. This maglev system, unlike the one previously tested at San Diego, would keep the area directly above the magnet platform free and provide the surgeon easier access to the patient, another pig or, eventually, humans.

In this chapter, we have seen how the use of sensors and feedback (and the ability of electromagnets to change magnetic fields by changing electric currents) allowed us to float magnet-containing objects on our desks, in wind tunnels, in plasmas, and even inside the abdomen of a pig. In some cases, like the floating globes and MIT's superdipole, the goal was simply to levitate the object in a stationary position, while in other cases, like the microbots and the maglev surgery platform, the objective was also to remotely control the motion of the levitated object. In the next two chapters, we shall see how these possibilities have been put to practical use in a wide variety of applications.

In a Spin

Bearing Up

As noted in the Preface, "Fighting the forces of gravity and friction is one of the things that magnets do best." The two fights are often very closely related, but we focus now on the fight against friction. Most machines have both moving parts and stationary parts, and friction occurs whenever the surface of a moving part is in contact with the surface of a stationary part. That friction can cause wear and abrasion of one or both surfaces, make motion less smooth, waste energy, and heat parts you really don't want to heat. One popular approach to fight friction is with the use of mechanical ball bearings or roller bearings (rolling spheres or cylinders)—it is much easier to roll things than to slide things. Another common approach is through lubrication with oils or other fluids, either alone or combined with ball bearings. But another approach that has become more and more popular in recent decades is the topic of this and the following chapter—*magnetic bearings*. With magnetic bearings, magnetic forces can reduce or even totally eliminate the direct contact between the moving and stationary parts. If you reduce the contact, you reduce the friction. If you are fully levitated and contact-free, you can be friction-free. No contact, no friction. No wear and abrasion, smoother motion, less wasted energy, less unwanted heating. Sounds almost like magic.

In this chapter we concentrate on *rotational* machines in which the moving part is a spinning *rotor*. (The stationary part is called the *stator.*) Magnetic bearings are commonly classified as *active* bearings if they use sensors and feedback control, and *passive* bearings if they do not. A very simple example of a passive magnetic bearing is the Revolution toy we discussed in Chapter 3 (Figure 4). Repulsion between circular permanent magnets on the rotor and triangular permanent magnets in the base (Figure 6) levitates the rotor and stabilizes four degrees of freedom, two translational and two rotational. Of the remaining two degrees of freedom, rotation about the axis of the rotor is free (a highly desirable property for a rotational machine), but the other one, translation along the axis of the rotor, is unstable. Consistent with the dictates of Reverend Earnshaw, we can't achieve full contact-free levitation with permanent magnets alone, but here we come pretty close. All we need is a point contact of one end of the rotor with a vertical glass plate. Although we haven't fully removed friction, we have limited it, and the Revolution can spin for several minutes before the modest friction of a rotating steel point against the glass surface slows the rotor to a stop.

The frictional force on a moving surface is proportional to the force exerted perpendicular to the surface, with the ratio between the two forces called the *coefficient of friction*. The coefficient of friction can vary from as low as 0.04 for steel on Teflon to more than 1 for metal on metal. In the Revolution, the rotor is horizontal and the perpendicular force between the tip of the rotor and the glass plate is a magnetic force, resulting from the Earnshaw instability in that direction. In many cases, however, the rotor is vertical, and the force between the rotor and the support structure below is its weight. If you can balance the rotor weight with an upward magnetic force, you reduce the force on the bottom support structure and thereby reduce friction. You fight friction directly by fighting gravity.

There are several useful and important rotational machines with vertical rotors that do not achieve full contact-free levitation, but use passive magnetic bearings to greatly decrease friction by balancing the rotor weight with upward magnetic forces. One of those machines you have at your house or apartment—your electric meter, or more precisely, your *watt-hour meter* (Figure 26). Watts are a measure of electrical power, the rate at which you use electrical energy, and watt-hours are a measure of the total energy you use, for which you will be billed each month. In my Massachusetts town, the rate is seven cents for 1,000 watt-hours, that is, a kilowatt-hour (kWh). In the watt-hour meter, the rotor axis is vertical, and upward repulsive forces between like poles of permanent magnets on the rotor and in the base compensate for most of the downward gravitational force on the rotor. Guide pins at the top and bottom provide horizontal stability, but with the magnetic forces nearly balancing gravity, very little friction and wear take place. That's important for a meter that is designed to last for many years with the rotor in continuous smooth and accurate rotation. In your watt-hour machine the rotational speed is quite slow, or at least I hope it is for the sake of your wallet. For most of the other rotors supported by magnetic bearings, the rotational speed is much higher.

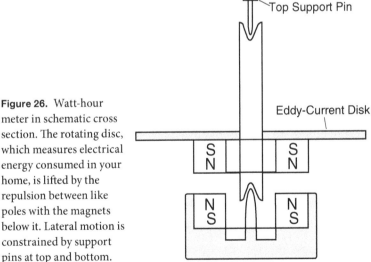

Figure 26. Watt-hour meter in schematic cross section. The rotating disc, which measures electrical energy consumed in your home, is lifted by the repulsion between like poles with the magnets below it. Lateral motion is constrained by support pins at top and bottom.

Enriching and Proliferating

You may have seen the pictures—Mahmoud Ahmadinejad smiling broadly as scientists and engineers in white lab coats lead the Iranian president past rows and rows of shiny vertical cylinders (Figure 27a). The occasion was his widely publicized April 2008 visit to Natanz, a now not-so-secret facility in central Iran dedicated to the enrichment of uranium. Iran claims the enriched uranium will be used only for peaceful energy production, but the West worries that it will be used for nuclear weapons, making those tall shiny cylinders one of the hottest issues in international relations. They contain centrifuges, machines using magnetic bearings to fight gravity and friction and allow their vertical rotors to spin at high speeds to enrich uranium. And they have a fascinating history, a history of international intrigue that leads from Germany to Russia to the Netherlands, and then to Pakistan, China, North Korea, Iran, and Libya. More than enough for several James Bond novels, but as physicist Jeremy Bernstein put it, the history of those centrifuges "is so implausible that if one put it in a novel, no one would believe it."

The story starts at Germany's surrender in 1945, when the Russians and Americans, then officially allies but realizing that they soon would be enemies, competed frantically for the spoils of war in the form of German scientists and engineers. One of those was Gernot Zippe, an Austrian with a degree in physics who had spent the war with the Luftwaffe. The Russians took him from a prisoner-of-war camp and moved him to Sukhumi, a city on the Black Sea that for Zippe and his fellow captives was something of a golden cage. It was a pleasant resort city, and they were treated well, but they were prisoners. (Today Sukhumi is the capital of Abkhazia, a disputed part of Georgia with allegiance to Russia). In Sukhumi the captured Germans were assigned the task of using various techniques to separate isotopes of uranium.

Natural uranium consists mostly of the U-238 isotope, with 92 protons and 146 neutrons in each nucleus. Less than one percent is

Figure 27. (a) Iranian president Maumoud Ahmadinejad walking with others between rows of centrifuges used to enrich uranium. (b) Ahmadinejad and others in front of a display of centrifuge parts. The large black object to the left of the shiny outer casing is the carbon-fiber rotor, and immediately to the left of the rotor can be seen two small rings. These are the magnetic bearings that support much of the rotor weight.

U-235, an isotope with three less neutrons, which is of interest for uranium fission bombs like the one that destroyed Hiroshima. To be useful for such bombs, you want to produce uranium enriched to as much as 90% U-235, and that isn't easy because the two isotopes differ in weight by only about 1%. Most of the uranium in the Hiroshima bomb had been enriched in hundreds of electromagnets at Oak Ridge, Tennessee. There the Manhattan Project used Fact 9, the magnetic force on moving charged particles, to separate U-235 and U-238 ions through the different paths they took in magnetic fields as a result of their slightly different masses. But the captured Germans at Sukhumi were trying other methods, and Zippe's group in particular was directed to try the centrifuge approach.

Centrifuges had been in use since the nineteenth century for separating things of different weight. Items in spinning machines are pushed to the outside, like clothes in the washing machine and people in spinning amusement-park rides, and heavier items become concentrated near the outside. The centrifuges used to enrich uranium were gas centrifuges, the uranium inserted in the form of gaseous uranium hexafluoride (a molecule with six fluorine atoms attached to each uranium atom). The faster the spin, the better the separation between heavy and light objects. To get faster spin and better separation of the gas molecules containing U-238 and U-235, Zippe and his colleagues decided to use a passive permanent-magnet bearing at the top of the rotor to balance much of the rotor's weight. Like the designers of the watt-hour meter, they fought friction by fighting gravity. The 2008 pictures of Ahmadinejad's visit to Natanz include one showing him admiring a display of the various components of the Iranian centrifuge, and among the visible parts is a pair of ring magnets that comprise the upper bearing of the rotor (Figure 27b). Centrifuge designs have advanced since the first ones built by Zippe in the 1940s, but they still seem to use permanent-magnet bearings. One ring magnet is presumably mounted at the top of the rotor and attracted upward to another ring magnet fixed to the support structure. (Here we should admit that such things as details of uranium centrifuge design remain deeply guarded secrets.) Iranian attempts in 2004 to procure several thousand ring magnets from black-market suppliers were among the clues that they were planning to build thousands of centrifuges.

As in the watt-hour meter, the rotor in the Zippe centrifuge is not fully contact-free and perhaps should be called magnetically *lifted* rather than magnetically levitated. The rotor is in contact at the base and spins on a needle bearing, but because much of the weight of the rotor is balanced by upward magnetic forces at the upper bearing, the needle bears little weight and friction is low.

The Zippe centrifuge used heating to set up temperature differences and convection movements in the gas that tended to move *down* the heavier molecules with U-238 that had been preferentially spun to the outside and, near the center, move *up* the lighter molecules containing U-235 and deliver them to scoops to collect the enriched gas. The enrichment from a single centrifuge is very modest, the concentration of U-235 going in the first step only from about 0.7% to about 0.9%, so many centrifuges in series are necessary to reach the concentration of 3 to 5% desired for power reactors. And many, many more spins are necessary to reach the concentrations approaching 90% needed for nuclear bombs, in case you are more interested in destroying cities than in lighting them. In November 2007, Ahmadinejad bragged that Iran had 3,000 centrifuges in operation, and in April 2009 he announced 7,000, with an ultimate goal of 50,000. But of course all the enriched uranium, he assured the world, would be used only for peaceful purposes.

The Zippe centrifuge moved from Russia in the 1940s to Iran decades later through a very circuitous route. When Zippe was released by the Russians and returned to the West in 1956, he was astonished to learn that the centrifuge that he had designed for the Soviet Union was better than anything in the West. It was the best in the world in fact. Soon he went to the University of Virginia and assisted in centrifuge designs for the West that eventually became the property of the transnational company URENCO, formed in 1970.

URENCO had a branch in the Netherlands, where in 1972 a Pakistani metallurgist named A. Q. Khan was hired and, since he was fluent in both German and Dutch, was given the task of translating the German centrifuge plans into Dutch. After India tested their "Smiling Buddha" nuclear weapon in 1974, Pakistan's president Zulifkar Ali Bhutto decided that Pakistan needed one as well and asked Pakistani scientists all over the world to return to their homeland to work on it. Khan wrote Bhutto offering his services, which included stealing plans and parts for the Zippe centrifuge and turning them over to Pakistan. Bhutto eagerly accepted and soon gave Khan his own nuclear facility near Islamabad.

Khan had centrifuges to enrich uranium, but Pakistan had no bomb. China had already built and tested a bomb, but lacked advanced centrifuges. In return for help with centrifuges, China helped Khan design the Pakistani bomb. Pakistan wanted missiles, and got missile designs and parts from North Korea in exchange for help with centrifuges. Khan expanded into the nuclear business in a big way, offering to sell centrifuges and other nuclear-related materials for cash to anyone interested, and among his eager customers were Iran and Libya. Starting in 1987, Iran bought centrifuge designs and parts from Khan, and the first uranium centrifuges assembled in Iran are called P-1 for Pakistan. Khan's deals with Libya, however, were his undoing. He began secretly delivering centrifuges and parts to Libya in 1997, but in 2003 a ship delivering material to Libya was intercepted and Khan's international nuclear network was finally exposed and shut down. Khan, still a national hero in Pakistan for bringing his country the bomb, became an international pariah for his contributions to nuclear proliferation. And it all started back in the 1940s with the Zippe centrifuges, their rotors spinning many thousands of revolutions per minute to enrich uranium with the help of their gravity-fighting and friction-fighting magnetic bearings.

A Few Words about Energy

I've referred to "energy" a few times in previous chapters. For example, at the end of Chapter 4 I noted that motors convert electrical energy into mechanical energy and that generators do the reverse, converting mechanical energy into electrical energy. I'll be talking about energy often in the next section, so it's worthwhile here to say a little about how scientists and engineers use the word. It overlaps a bit with how nonscientists use the word, but when we scientists and engineers speak of energy, as we do a lot, we are often referring to something quite specific and quantifiable in terms of numbers and/or equations.

We talk about energy associated with objects in motion, and we call that *kinetic energy,* a form of mechanical energy. We of course have an equation relating kinetic energy to the velocity and mass of the moving object, but we won't need it here. The unit for energy most commonly used by physicists is the joule, which is one watt-second. (The watt is a measure of power, i.e., the rate at which energy is generated or used.) The electrical industry prefers to quote energies in kilowatt-hours, which is a lot bigger unit. One kilowatt-hour equals 3.6 million joules.

We also talk about a more abstract concept of energy, *potential energy,* for situations in which an object has the "potential" for doing something useful, like producing kinetic energy. There are many different types of potential energy, and the most familiar is *gravitational* potential energy. If you lift an apple off the ground, we say that you are increasing its gravitational potential energy. If you then let it go, the apple falls faster and faster, converting that gravitational potential energy into the kinetic energy of falling.

And that form of energy conversion can go both ways. On a swing, your gravitational potential energy is at a maximum when you are at the top of your swing. As you swing down, you lose potential energy and pick up kinetic energy until you reach the bottom of your swing, when your kinetic energy is at a maximum. But as your momentum carries you back up again, that kinetic energy converts back to gravitational potential energy. You are losing a bit of total energy through friction, so you might not swing up quite as high as you were before unless you add a bit of your own muscle energy to pump yourself up.

Another form of potential energy important to the topics we discuss in this book is *magnetic* potential energy. If you have a small ball of iron stuck on a magnet and you pull it away from the magnet, you are increasing its magnetic potential energy. If you then let it go, it will be pulled back to the magnet, moving faster and faster, converting magnetic potential energy into kinetic energy. When it hits the magnet, it will convert part of that kinetic energy into sound energy and the rest into internal mechanical stresses in both the magnet and the iron ball. With a brittle neodymium magnet, those internal stresses can sometimes even break the magnet, as some have learned to their dismay.

A compass needle in the magnetic field of the earth has minimum magnetic potential energy when it lies parallel to the earth's field. If you turn the compass and the needle at first turns with it, the needle's magnetic energy increases, and it will swing back—magnetic potential energy converting into kinetic energy of rotation—to point north-south, where it has minimum magnetic potential energy.

When allowed to, objects will decrease their potential energy. If you let them go, the apple will fall, the iron ball will move to the magnet, and the compass needle will turn to point with the field. But what about that top ring magnet in Figure 3? It feels the competing claims of gravitational and magnetic potential energy, a competition basic to magnetic levitation. The magnet can decrease its gravitational potential energy by falling, but the ring magnet below it has a like pole repelling it upward, so coming closer *increases* its magnetic potential energy. At the stable separation distance where the downward gravitational force equals the upward magnetic force, the magnet's *total* potential energy, gravitational plus magnetic, is at a minimum. For an object free to move, *stable equilibrium corre-*

sponds to minimum potential energy. Any movement from that position will increase total potential energy and produce restoring forces. Thinking in terms of gravity, it's in an energy valley.

Back in Chapter 3, we also discussed a case of *unstable* equilibrium, where at a particular position of a magnet, an *attractive* magnetic force exactly balanced a gravitational force but displacements from that position led not to restoring forces, as in the stable repulsive case, but to net forces that tended to move the magnet *away* from that position, that is, to destabilizing forces. Stable equilibrium corresponds to minimum potential energy, but *unstable* equilibrium corresponds to *maximum* potential energy. As with the floating globe on my desk, there you need *feedback control* (or diamagnetic barriers, as in Figure 15 or 16) to keep an object in position and to resist its natural tendency to decrease its potential energy by falling off the energy hill.

For some displacements, like rotating the Revolution toy, a watthour meter, or a uranium centrifuge around its axis, there is neither stability nor instability, but neutral stability. The potential energy is unchanged by rotation, with no minimum or maximum, no energy valley or energy hill, no net restoring forces or destabilizing forces. The rotor can just turn, turn, turn.

We remember from the Reverend Earnshaw that we can't achieve full stability with just the attractive and repulsive forces of static magnets. Translating Earnshaw's rule into energy terms, we can't have a position for the magnet where you have an absolute energy minimum, a position where energy increases for all possible displacements. There will always be at least one displacement that results in a decrease in energy, a direction in which the magnet is unstable. One permanent magnet above another in *repulsion,* like the ring magnet on the pencil in Figure 3, is stable vertically but unstable horizontally, so it needs the pencil to hold it there. A permanent magnet below another in *attraction* is stable horizontally but unstable vertically (as we noted again two paragraphs ago). We call such situations, where there's an energy minimum in one direction and an energy maximum in another direction, a *saddle point,* since a horse's saddle commonly curves one way along the horse, a minimum, and the other way across the horse, a maximum. If the horse is stationary, you're more likely to fall off sideways than forward or back. This may be more than you wanted to know about potential energy and stability, and I won't refer to saddle points again in this book. But I think that saddle point is a very cool term that people have invented to help them visualize a somewhat complex energy concept, and I thought it would be fun to include it.

In Chapter 3, we discussed stable and unstable equilibrium in terms of competing gravitational and magnetic *forces.* Here we discussed them in terms of competing gravitational and magnetic *energies.* But the two seemingly different descriptions are equivalent, since energies and forces are directly related to each other. We've chosen to avoid mathematical equations in this book, but forces are equivalent to *energy gradients,* changes of potential energy with position. That's an important basic concept that you can file away to impress people at your next cocktail party, but let's now leave this digression into abstract scientific terminology and return to the real engineering world of magnetic bearings and spinning machinery.

Flying Wheels

The rotors of watt-hour meters and Zippe centrifuges, supported only by forces between permanent magnets, were not fully contact-free. To achieve fully contact-free rotors, most magnetic bearings are active bearings using sensors and feedback control like the floating globes and other devices of the preceding chapter. For one example, consider the spinning rotors of flywheels (sometimes called momentum wheels) used in energy storage.

Energy storage has always been of interest to the electrical power industry because of the strong variations in demand for power— differences between days and nights, seasonal weather changes, and other causes. In the future, as we increase the amount of power from renewable sources like solar and wind, the intermittent nature of such sources will further increase the need for energy storage systems in the electrical grid. There are also short-term needs for energy storage. Some customers, for example, have operations that would suffer from power outages of even a few seconds, and require what is called Uninterruptable Power Supply (UPS)—a system that requires stored energy ready and available to fill in until the outage is over or until backup power is operational.

One low-tech solution for *long-term* needs for balancing supply and demand for power has been hydropumping—using excess power at low-demand times to pump water to higher elevations, storing electrical energy in the form of gravitational energy, and releasing the water at high-demand times to generate electricity. Another approach to energy storage is with batteries, which store energy chemically. But we are currently interested in rotating machines, so we focus on *flywheels,* which store energy in the form of mechanical energy, that is, in the kinetic energy of rapidly spinning rotors. For energy-storage applications, they are sometimes called mechanical batteries. Flywheels are attached to a motor-generator that acts as a motor for energy input, converting electrical energy into mechanical energy by accelerating the flywheel, and acts as a generator for energy output, slowing the flywheel and converting the mechanical energy back into electrical energy when needed. For applications like this, which require many thousands of revolutions per minute (rpm) in rotation, it is desirable to have very low friction, a goal that is often met by operating in vacuum to avoid air drag and by supporting the rotor with magnetic bearings, which are not only low friction, but also more favorable for operation in vacuum than lubricated mechanical bearings.

One of the leaders in flywheels for very short-time energy storage in UPS systems is California's Pentadyne Power. They started supplying flywheels to customers in 2003 and within five years had several hundred flywheels in commercial use around the country. Like the watt-hour meter and the Zippe centrifuge, their flywheels rotate about a vertical axis, but the flywheel rotors spin *fully contact-free*. In the cabinet containing the electronic controls for the flywheel is a "magnetic levitation module" containing five separate feedback circuits using signals from position sensors to control five degrees of freedom of the spinning rotor. The sensors that detect the rotor position in the Pentadyne system are *capacitive sensors*. A capacitor is a circuit element that essentially consists of two metallic conducting surfaces separated by a small insulating (nonconducting) gap. Like the inductance of an electromagnet coil, the *capacitance* of a capacitor produces a complex frequency-dependent relation between currents and voltages in an AC circuit. And the capacitance is very sensitive to the width of the insulating gap between the two conducting plates, which makes it a very sensitive gap sensor and position sensor. Changes in that gap produce electrical signals that the feedback circuits translate into signals to the electromagnets to adjust currents and magnetic fields to control the rotor position. A diagram on the Pentadyne website shows an "axial electromagnet" at the top of the rotor to provide the magnetic force that controls vertical displacement, and "radial electromagnets" at the top and bottom to provide magnetic forces to control lateral displacements. By fighting gravity with magnetic levitation, they have greatly decreased friction, resulting in a system that requires less energy to run and less maintenance than flywheel systems using mechanical bearings.

The total weight of their rotor is only 52 pounds (24 kg), half of which is the central shaft and the rotor part of the motor-generator that accelerates or decelerates the flywheel to accept or deliver power, while the other half is the flywheel itself, a carbon-fiber cylinder connected to the shaft by a titanium hub. Composite materials strengthened by carbon fibers are much stronger than the steels and alloys used for flywheel rotors in the past. This allows the Pentadyne flywheel to spin at speeds greater than 50,000 rpm and sustain the associated huge outward forces without bursting. Designed to combat power outages of several seconds, the system can deliver 190 kilowatts (kW) of power for about 10 seconds. The stored energy in their flywheel at maximum rotational speed is 0.74 kilowatt-hours (kWh). Their systems have been installed in numerous hospitals and hundreds of missile silos around the country. It's comforting to know that our ability to obliterate our enemies with nuclear weapons will not be curtailed by a brief power outage.

Beacon Power of Massachusetts offers flywheels with a slightly lower power rating (100 kW) but much greater total stored energy (25 kWh). The kinetic energy stored in a spinning flywheel is proportional to its mass and to the square of its rotational speed. Although the maximum rotational speed of the Beacon flywheel is 16,000 rpm, considerably less than that of the Pentadyne flywheel, their flywheel mass is over a ton, yielding greater stored energy for longer-term energy storage needs. Their system uses a passive permanent magnet bearing and an active magnetic bearing, using inductive sensing, to lift most of the rotor weight, but also has small ball-bearing contacts at top and bottom. Although their spinning rotor is therefore not completely contact-free, the friction is nevertheless remarkably low for such a heavy rotor. A group of ten Beacon flywheels, with a total power rating of one megawatt (a million watts) was installed in the New England power grid in the fall of 2008 to assist in power regulation. By 2010, Beacon had three 20 megawatt flywheel assemblies under development. There appears to be growing interest in flywheel systems for both short-term and long-term energy storage.

Another interesting recent development in maglev flywheels is the construction of systems using passive superconducting bearings. The discovery of high-temperature superconductors in the late 1980s stimulated interest in their use in passive magnetic bearings, and early work in this field was summarized by Francis Moon in *Superconducting Levitation* (1994). Recently, Boeing described a 5 kWh/100 kW flywheel with a superconducting bearing built from YBCO (yttrium–barium–copper–oxygen). Their system uses a passive permanent-magnet bearing to lift most of the rotor weight.

This bearing by itself is stable for radial (horizontal) displacements but axially (vertically) unstable, while the superconducting bearing provides the axial stability and additional radial stability. It's a totally passive system, with no need for sensors and feedback circuits. The superconducting bearing consists of an assembly of many hexagonal pieces of YBCO, with strong flux pinning, repelling upward a rotor containing neodymium magnets. In early designs the YBCO was immersed in liquid nitrogen, but more recently it is conduction cooled by intimate contact with a nitrogen-cooled copper plate. It is "field cooled" like the YBCO discussed in Chapter 7—cooled to below its superconducting critical temperature in the magnetic field from the rotor. Boeing reports extremely low frictional energy loss, even when including the energy associated with the cooling. A 5 kWh/250 kW flywheel using YBCO bearings was recently reported by ATZ, a German company, which noted that only about 10 pounds of YBCO levitated a rotor weighing more than a ton.

Pumping Blood

One machine with a magnetically levitated rotor that may some day save your life or mine is a blood pump, or ventricular assist device (VAD). VADs without magnetic bearings have been in use for about four decades now, but the new maglev or "third-generation" devices have entered into clinical trials and, with their rotors spinning contact-free without friction and wear, are expected to be capable of many years of service to relieve, or eventually replace, tired hearts. There are also claims that maglev heart pumps are potentially less damaging to blood cells than traditional pumps.

If your heart is healthy, its left ventricle pumps blood through your aorta to your body's thirsty cells at a rate of about 5 liters (quarts) a minute—more when you are exercising. After the oxygen-depleted blood returns to your heart, its right ventricle pumps it to the lungs to pick up oxygen, and the freshly oxygenated blood returns to your heart to start the process all over again. But if your heart muscle is damaged and unable to pump enough blood to keep you going, an implanted VAD can fill in. Normally it does the work of the muscle of the left ventricle, taking blood from there and delivering it to the aorta. For some patients, just resting the heart muscle for a couple of months in this way allows the muscle to recover, and the VAD can be removed ("explanted"). In other cases, the VAD can fill in for the ventricle until a heart transplant becomes available. More than 2,000 heart transplants are performed every year in the United States, but that's not nearly enough for all the patients who need them. So there is hope that eventually such "artificial heart" devices will become simple enough and reliable enough to keep patients alive, ambulatory, and active for many years without the need for a transplant. Using VADs for a limited time until the heart recovers or until a transplant is available is called bridge therapy, while eventual long-term use is known as destination therapy.

Early VADs pulsed like our hearts, but most VADs used today, including the maglev blood pumps now in clinical trials, provide continuous blood flow. In the maglev pumps, the rotor or impeller is the only moving part, and it spins fully contact-free, driven and stably levitated by magnetic fields from electromagnets in the housing. Although rotors in most machines spin about either vertical or horizontal axes, rotors in blood pumps must of course accommodate to changing orientations when the patient moves, as when a patient who was lying in bed sits up or stands up. The magnetic forces on the rotor in these devices are far stronger than gravitational forces, so changes in the direction of gravity require only small adjustments in currents to keep the rotor in its stable position.

The first patient in the United States to receive a maglev blood pump was Anthony Shannon, a 62-year-old from Livonia, Michigan, whose heart had been weakening for twenty years after a heart attack and clogged arteries had damaged his heart muscle. On July 30, 2008, at the University of Michigan Cardiovascular Center in

Ann Arbor, a DuraHeart VAD was implanted in his chest. Although Shannon was the first patient in the United States to receive a Duraheart, he was the 71st in the world. Clinical trials of the Duraheart started in Europe in 2004, and after several years of successful experience with advanced heart failure patients, one living with the implant for over three years, trials were finally approved in the United States. The DuraHeart (Figure 28) is about the size of a hockey puck and is a *centrifugal pump,* with the spinning, levitated, disc-shaped rotor pumping blood radially outward. An electrical lead through the skin connects to a controller and battery pack that can be worn in a carrying bag attached to the belt. The Levacor, a competitive centrifugal pump from a different manufacturer, has also had clinical trials in Europe. The first implant of a Levacor in the United States took place in January 2010 in Oklahoma City. A maglev heart pump of different design that has seen considerable clinical testing in Europe is the Incor, a blood pump with a rotor shaped like a metal screw that pumps blood along its axis; that is, it is an *axial pump* rather than a centrifugal pump. Early results with maglev heart pumps like the Duraheart, Levacor, and Incor are promising, and others are being developed, in various stages of the lab tests, animal tests, and clinical trials on humans required before they are approved for general use. It remains to be seen whether maglev blood pumps will eventually replace the VADs in most common use today.

One of the claimed advantages of maglev blood pumps is long life, resulting from reduced friction and wear on the rotor and housing, but improved durability is of course one advantage that will require many years to establish. It's comforting to know that blood pumps will be getting better and better in the coming decades, but I hope that with a good diet, plenty of exercise, and a generally healthy lifestyle, you and I can keep our hearts strong enough that they won't need mechanical help, maglev or not.

Figure 28. Exploded view of DuraHeart maglev blood pump. With active magnetic bearings, the rotor (center) spins completely contact-free of its housing. Photo courtesy of Terumo Heart Inc.

Floating Rotors

Hundreds of other types of rotating machines use magnetic bearings to produce rotors that spin contact-free with reduced friction and noise, improved energy efficiency, smoother and faster rotation (rotational speeds up to over 100,000 rpm), longer life, and reduced maintenance, repair, and pollution compared to oil-lubricated mechanical bearings. They include high-speed *motors* that convert electrical energy to mechanical energy and *generators* that do the reverse. They include *turbines* that convert the pressure of moving liquids and gases on their blades into spin and *pumps* that use spinning blades to move liquids and gases. They include "turbo-molecular pumps" that remove gas molecules from containers to produce high vacuum. Magnetic bearings are especially popular for turbo-molecular pumps because they avoid the use of lubricants from which molecules can evaporate and limit achievable vacuums. Magnetic bearings are even used nowadays in some cooling fans. Maglev fans are quieter than most fans and are used in some laptop computers.

In the 1950s, Jesse Beams of the University of Virginia pioneered the development of active magnetic bearings and built ultracentrifuges that spun small rotors at millions of rpm. Later NASA developed active magnetic bearings for gyroscopes and flywheels used to control the "attitude" (orientation) of satellites. A related military application is in inertial-guidance systems for missiles. One of the first large-scale applications of magnetic bearings in industry was in compressor pumps in natural gas pipelines. Natural gas is transferred hundreds of miles through pipes under high pressure, and every hundred miles or so there is a compressor station. Replacement of machines using oil-lubricated mechanical bearings by machines with magnetic bearings both saved substantial operating energy and removed various problems associated with the lubricating oil, including contamination of the natural gas, disposal of used oil, and fire hazards. Speaking of oil (of a different kind), machines supported by magnetic bearings are also commonly used to pump oil through oil pipelines.

In 1996, Nippon Koei, a Japanese power company, put into operation a water turbine and power generator with a vertical rotor weighing 14 tons centered by three radial magnetic bearings and supported by one axial magnetic bearing. At the time, it was reportedly the heaviest rotor supported by magnetic bearings. As with many large machines levitated with magnetic bearings, there were also auxiliary or "touchdown" mechanical bearings to support the rotor in case an outage turned off the power to the active magnetic bearings. (Unless you have an Uninterruptible Power Supply system, active bearings generally require a fail-safe backup in case of power failure.) The change from oil-lubricated mechanical bearings to magnetic bearings was reported both to increase the energy efficiency of the system and to eliminate oil pollution of the river. More recently, Nippon Koei has installed even larger and heavier water turbines and generators that use three radial bearings to center a spinning shaft 7 meters (yards) long and an axial magnetic bearing more than a meter in diameter capable of sustaining a total downward force of up to 35 tons. The monster bearings in that machine were provided by S2M, a French company that in 1976 was the first company to market active magnetic bearings, originally for aeronautical and space research. In those Nippon Koei turbines, the S2M magnets "bearing up" as much as 35 tons of load really do some heavy lifting!

For machines like the huge Nippon Koei turbines that rotate about a vertical axis, it is the axial (thrust) bearings that provide the antigravity force, whether it requires a few pounds or many tons. Many other machines rotate about a horizontal axis, and here it is the radial bearings that do most of the work of fighting gravity. Whatever the orientation of the rotation axis may be, radial bearings commonly contain on the stator a series of electromagnets surrounding the steel shaft of the rotor, and the various electromagnets, following signals from sensors (usually inductive or capacitive) and feedback circuits, provide a balance of forces to oppose off-center displacements of the rotor. To resist tilting of the rotor shaft, it is often held and centered at two separate points with two radial bearings.

Some manufacturers of rotating machinery based on magnetic bearings call their machines "bearingless," since they lack traditional mechanical bearings and their rotors spin contact-free. For example, a Swiss company called Levitronix (despite its name, it has no connection to the Levitron toy) markets "bearingless motors" and "bearingless pumps," including blood pumps. Their motors and pumps typically have two radial magnetic bearings, each one containing four electromagnets, and one thrust bearing with two electromagnets, for a total of ten electromagnets. Stable levitation of the rotor in five degrees of freedom is provided by five inductive sensors and feedback circuits that control the currents to the electromagnets. For both mechanical bearings and the magnetic bearings in "bearingless" machines, the properties of practical interest include the total load capacity (related to the weight of the rotor and forces exerted on it in operation), the "stiffness" (the strength of restoring forces resisting small displacements), damping of vibrations, size (magnetic bearings are often bigger than equivalent mechanical bearings), and power consumption. And one of the most important properties of any engineering device—price.

Since contact-free and friction-free magnetic bearings are usually more expensive than traditional mechanical bearings, their use has to be justified to purchasers by the various improvements in performance noted earlier, but nowadays also including claims that magnetic bearings are more environmentally friendly, more "green." In that regard, improved energy efficiency and longer operating life and decreased maintenance are definite pluses, as is decreased pollution from lubricating oils. Marketers of maglev flywheels for energy storage emphasize how green they are compared to their major competitors, that is, batteries containing all those nasty chemicals that pose a serious disposal problem. One maglev machine that had instant appeal to green enthusiasts is the maglev wind turbine or "wind power generator," first announced in China in 2006. The low friction attained with magnetic levitation was claimed to make their turbine capable of generating power at much lower wind speeds than earlier wind turbines, producing some useful electrical power even in a mild breeze. It is marketed in China by ZK Energy, Hunan, and at the time of writing, their website offers only 300 watt and 600 watt models of a "full permanent magnetic suspension wind power generator," but promises higher-power models soon.

Like most current wind turbines, and the ancient windmills that preceded them, the blades of the ZK Energy machine rotate about a horizontal axis. However, Enviro-Energies of Canada and the United States offers maglev wind turbines that rotate about a vertical axis and are designed to mount on your rooftop (Figure 29)—perhaps alongside your solar panels, to show your neighbors just how environmentally conscious you are. Their turbine blades look more like curved metal sails than propellers, and unlike turbines with a horizontal axis of rotation, vertical-axis turbines don't need to be turned to face the wind. Enviro-Energies offers models rated from 2.5 kW to 10 kW, and for a given power rating, the size of the turbine blades you'll need depends on your local wind conditions—less wind, bigger blades. The wind speed required to reach peak power depends sensitively on the shape and orientation of your roof, which, if you're lucky, can make the wind speed at the turbine significantly higher than the wind speed down the street. As a result, even winds as low as 10 mph (16 km per hour) can sometimes generate appreciable power. One Arizona company claims to have plans for a maglev wind turbine that will generate up to 2 gigawatts (2 billion watts) of power, but that turbine will be too big to put on your roof. Their online video announces, however, that it will require less than 100 acres of land. They are thinking big. A representative of that company made a presentation at a 2008 maglev conference on their large wind turbines, and his very appropriate name—I kid you not—was Larry Blow.

As in watt-hour meters and Zippe centrifuges, the spinning rotor in the Enviro-Energies "maglev" wind turbines is not completely contact-free. The shaft has light mechanical contact to provide lateral stability, but passive permanent-magnet bearings provide upward

Figure 29. Rooftop "maglev" Enviro-Energies vertical-axis wind turbine, rated at 5 kW.

repulsive forces to carry most of the weight, thus yielding very low friction. And since the spinning blades mostly float on magnetic fields, they transmit little noise or vibration to the roof. For another plus, over time the magnetic bearings are likely to need less maintenance and repair than mechanical bearings. A set of permanent magnets rotates with the turbine, inducing eddy currents and generating electrical power in stationary coils below through relative motion (Chapter 5). Ed Begley Jr., Hollywood actor and environmental activist, has become a spokesman for this company, and a recent video session of "Jay Leno's Garage" featured Leno admiring one of these rooftop turbines and vowing that he would install one on his house bigger than the one on Begley's house. When celebrities as prominent as Jay Leno and Ed Begley Jr. promote magnetic levitation, perhaps maglev has finally arrived.

The No-Spin Zone

Flying Broomsticks

As we saw in the previous chapter, with the help of magnetic anti-gravity forces rotors spin with greatly reduced friction in watt-hour meters, centrifuges, flywheels, blood pumps, water and wind turbines, and many other rotating machines. In this chapter and the two following ones, we discuss some of the many *nonspinning* applications in which maglev also plays an important role. So be appropriately warned. With due apologies to Bill O'Reilly and the late Rod Serling, you have now entered another dimension. You are now in the No-Spin Zone.

Harry Potter and the other students at Hogwarts traveled through the air—and even competed in the challenging game of Quidditch—on flying broomsticks. With what form of magic those broomsticks were able to fly remained something of a mystery. However, when broomsticks were flying down a unique production line in Birmingham, Alabama, it was less of a mystery. They were flying with the magic of magnetic levitation.

The production line was the bold idea of Conrad Smith, founder of American Metal Handle, aided by David Trumper of MIT. Input to the production line was a roll of unpainted steel sheet, and the intended output was a series of painted steel tubes for broom handles. While on the line, in sequence, the sheet was formed into a tube, welded shut, sprayed with a coating of paint powder, induction heated to melt and bond the paint to the tube surface, cooled with a water spray, and pulled to a cutoff punch that sliced the painted steel tube into the desired lengths. Between the powder spray and the water spray, the traveling tube was supported only with magnetic forces.

Smith's production line was no small operation. The levitated portion of the traveling tube was over 50 yards (meters) long, with ten separate "suspension stations" spaced about 4 yards apart (Figure 30). Each such station had AC induction sensors to determine the position of the tube, feedback circuits, and electromagnet "actuators" to apply magnetic restoring forces to keep the moving tubes

Figure 30. Portion of the broom handle production line at American Metal Handle. On each side of the long induction furnace at the center are the first and second of the magnetic levitation stages. Metal tube can be seen at the right leaving the first levitation stage to enter the furnace, and at the left leaving the furnace to enter the second levitation stage.

in the desired position. An important problem encountered with such long contact-free structures is how to damp vibrations. To analyze and tackle this vibration problem, Trumper and his MIT students later constructed a much-reduced version of Smith's production line, with a suspended 3-yard long quarter-inch (6 mm) diameter tube (nontraveling) and multiple sensors and actuators. They found that electronic circuits averaging the output signals from two or more sensors and averaging the input currents to two or more actuators were helpful in damping vibrations in such systems; they felt their approach could be helpful in other noncontact lines for coating or painting, a general category of processing they termed levitated continua. But in Alabama, serious problems in the production line with the welding step, the powder-spraying step, and other parts of the automated production line eventually killed Smith's ambitious project to paint broom handles on the fly. With the help of electronic control systems designed to limit vibrations, the maglev portion of the production line worked well, but the system as a whole failed. As magical as it is, magnetic levitation unfortunately cannot solve all the problems of the world.

Flotors and the Maglev Touch

When I first learned of the maglev *haptic* program headed by Ralph Hollis of the Robotics Institute of Carnegie Mellon University, I confess I had to look up the word "haptic." My dictionary defines it as "related to or based on the sense of touch." My 1989 multivolume *Encyclopedia Britannica,* which I still tend to rely on for topics and events prior to its publication date, has no entry under haptic, but the online Wikipedia has several, including a lengthy discussion of *haptic technology.* Although the word "haptic" predates computers, clearly haptic technology has blossomed in the computer age. (Surprisingly, the spell-check function of my Microsoft Word does not recognize haptic as a word, but perhaps it will when I upgrade to a more recent version.)

Among our traditional five senses, touch is unique. First, it is not limited to one local organ, but is distributed about the body. Second, the hand, the focus of today's haptics technology, is a two-way device, capable of both input and output. Our eyes, ears, nose, and tongue can sense our environment by seeing, hearing, smelling, and tasting, but our hand not only can *feel* the shape and texture of a sculpture, it can *form* a sculpture. It can both sense our environment and change it. Some parts of haptic technology focus primarily on the *output* function of the hand, like the touch screen of the iPhone and like CNN's "magic wall" used for displaying and analyzing election results. Such systems *respond* to your touch, but usually don't deliver complex sensory input to your hand. In Hollis's maglev haptic system and the mechanical haptic systems that preceded it, however, the sensory *input* to the hand is equally important, if not more so. The systems not only respond to motions of the hand, they also feed back sensory signals to the hand. You *feel* things. About the maglev haptic system, Hollis states, "We believe this device provides the most realistic sense of touch of any haptic interface in the world today."

In public displays of the maglev haptics system, visitors could move the handle and maneuver, on a computer screen, an object within a transparent rectangular shape. You could move the object through all six degrees of freedom, three translational and three rotational, but when it hit the boundaries of the transparent container within which it was moving on the screen, you would realistically feel through the handle, in the real world, the surfaces, edges, and corners of the transparent container in the virtual world of the computer screen. In another display, you could use the handle to move a small object on the computer screen across surfaces of various textures (rough, rippled, smooth) in the virtual world, and through the real-world handle, realistically feel their textures. In another impressive display, you could feel the contours of a virtual rabbit! (Hollis brought his system to a 2010 conference in the Boston area, and I was able to experience these various demonstrations myself.) The convincing sensation of touch was achieved through the wonders of computer programming and magnetic levitation. Since the point of most maglev is to *avoid* actual touch and approach contact-free levitation, it is fascinating that maglev has also actually been used to *simulate* touch.

Other haptic systems link hand motions to the computer and return touch sensations to the hand through the motion of mechanical linkages, but mechanical systems suffer from static friction and associated irregular and irreversible disturbances to the motion of various important parts. In the maglev system, the central part, called the *flotor,* floats freely on magnetic fields, without mechanical contact with its surroundings except for several electrical wires dangling loosely below it (Figure 31). Rotors spin, but flotors only float; we are in the No-Spin Zone. In the shape of a hemispherical shell 5 millimeters thick (less than a quarter-inch) and with a radius of about 115 millimeters (4.5 in.), the thin flotor contains six current-carrying coils of wire and weighs 600 grams (1.3 lb) It requires about 4.5 watts to levitate in the field of two arrays of neodymium permanent magnets, twelve on the inner part and twelve on the outer part of the surrounding stator within which the flotor levitates and moves. The position of the flotor is constantly monitored with optical sensors, with light beams from three light-emitting diodes (LEDs) on the flotor imaged on position-sensing photodetectors (devices that convert the light image into electrical signals) on the stator. The handle is rigidly attached to the flotor, and changes in any of the six degrees of freedom of the handle and flotor are detected and communicated to the electronic control system and computer.

It sounds complex, and it is, but it works. The flotor and stator are housed in a spherical enclosure mounted in a tabletop surface, and the device, with its protruding handle, can be re-oriented by the user. (The control system and power supplies are located in a separate enclosure connected to the device by cables). Most of those who have used the maglev haptic system subjectively agree that it conveys a more natural sense of touch than the prior mechanical systems.

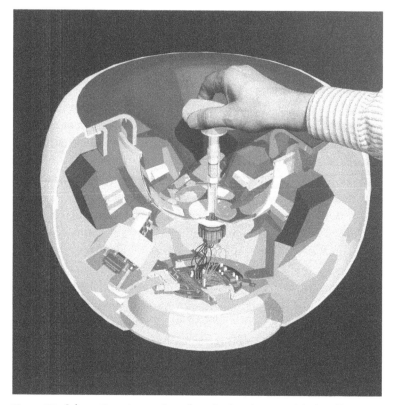

Figure 31. Schematic cutaway view of maglev haptics stage. The handle is rigidly attached to the thin flotor, which floats between two arrays of thick permanent magnets. The wires hanging loosely from the bottom of the flotor control the currents in the flotor coils.

There are also several objective quantitative measurements that agree, including increased stiffness (forces resisting displacements) and increased resolution (detection of finer surface textures). The major weakness of the maglev haptic system in comparison with its mechanical competitors is that, at least in its present form, it offers less range of motion of the handle. Because the flotor motion is confined to within the inner and outer parts of the stator, it and the attached handle can translate only about an inch in each direction, and rotate only about plus or minus eight degrees around each axis.

The forces that levitate the flotor result from the interaction between the fields of the two dozen permanent magnets on the inner and outer stator and the electrical currents flowing in the six separate coils on the flotor. Fact 10 of the basic "facts about the force" presented in Chapter 2 states that a current-carrying wire in a magnetic field experiences a force perpendicular to both the current and field. In the flotor, the magnetic field from the permanent magnets is constant, but the forces on the flotor (and the handle to which it is connected) are controlled by changes in the currents flowing in its six separate coils. When your hand moves the handle (and thus the flotor), the system changes the six currents to keep the flotor floating, and when the system returns sensory signals to your hand, it does so via changes in these currents. In physics courses, this force between magnetic fields and currents is presented in the form of an equation and is called the *Lorentz Force,* named after Henrik Lorentz, an outstanding Dutch physicist. His work not only impacted our understanding of electromagnetism, but also, importantly, he laid the groundwork for Einstein's special theory of relativity. Hollis and his colleagues refer to their method of levitation, based on Fact 10 and the levitation of current-carrying coils (air-core electromagnets), as Lorentz levitation.

Numerous universities and several commercial customers are now investigating possible applications of maglev haptics, including sensitive control of remote robot arms and medical and dental training systems. Haptic technology today still focuses on the hand as sender and receiver of signals. But, as noted earlier, our sense of touch is not limited to the hand. Perhaps some day haptic technology will expand to the extent imagined by Aldous Huxley in his futuristic 1932 novel, *Brave New World.* There he introduced the Feelies, public entertainments that included touch among the various sensations delivered by 3-D movies (which also were accompanied by a scent organ, so that the sense of smell was not neglected).

"Going to the Feelies this evening?" one character asked another. "I hear the new one at the Alhambra is first-rate. There's a love scene on a bearskin rug; they say it's marvelous. Every hair of the bear is reproduced. The most amazing tactual effects." In the theater, you had to take hold of metal knobs on the arms of your chair to get the Feely effects. When characters kissed on the screen, "the facial erogenous zones of the six thousand spectators in the Alhambra tingled with almost intolerable galvanic pleasure." But of course not all touch sensations are pleasurable. Later in the Feelie, a character fell on his head, and "a chorus of ow's and aie's went up from the audience." Haptic technology, even maglev haptics, hasn't progressed as far as the Feelies yet, but when they do, I hope they'll concentrate on the pleasurable aspects of the sense of touch, and not on pain.

Maglev Nanotechnology

Maglev haptic systems depend on one of the major marvels of the modern age—an electronic computer. Recently, magnetic levitation has come to play an important role in the manufacture of something very basic to the workings of electronic computers, as well as cell phones and all our other electronic wonders of today—the *integrated circuit.*

An important precursor to the integrated circuit was the *transistor,* invented at Bell Laboratories and announced in 1947. Transistors are based on semiconductors, materials like germanium and silicon that have much less electrical conductivity than metals, but features that make that conductivity more amenable to easy change and control. Transistors can switch electrical currents off and on, amplify electrical signals, and perform other electronic functions of their predecessors, vacuum tubes (which looked more or less like light bulbs). Although huge compared to transistors of today, even the earliest transistors were significantly smaller, cooler, faster, and cheaper than vacuum tubes, and they rapidly led to miniaturization of electronic devices from radios to televisions and computers. During my graduate school years in the early 1950s, I had a summer job experimenting with electronic circuits built with those new-fangled gadgets, which in those days were pea-sized and had three metal wires sticking out of them. I destroyed many transistors that summer, partly because the early ones were a bit delicate, but also because I really wasn't that good with electronic circuits. But I didn't destroy their ability to revolutionize electronics.

Transistors encouraged rapid growth in electronics, but as circuits grew more complex and involved more and more components, it became very tedious to assemble and wire together the many individual transistors, resistors, capacitors, and other parts of electronic circuits. In the summer of 1958, Jack Kilby, a new hire at Texas Instruments, had the idea of processing a single piece of semiconductor in such a way that several components of an electronic circuit could be *integrated* in their manufacture, that is, be made together on a piece of semiconductor without the need for later assembly. It worked, and in early 1959 he applied for a patent. By then another engineer, Bob Noyce, was working on a similar idea at Fairchild Electronics, and his integrated circuit, developed independently of Kilby, also worked. In some ways, it actually worked better, and Noyce was granted his patent in April 1961, while Kilby's patent application was still under consideration. Like Faraday and Henry and their independent discovery of electromagnetic induction, Kilby and Noyce were independent inventors of the integrated circuit. When Kilby received the Nobel Prize in Physics for it in 2000, Noyce surely would have shared it with him, but he had died ten years earlier, and Nobel Prizes are not awarded posthumously. (Long before, Noyce had quit Fairchild and joined two others to form Intel, a company that became a world leader in integrated-circuit "chips.")

The first integrated circuits contained only a few transistors, but later ones held thousands, then millions, and today they hold billions. This progress in numbers has been achieved not by going to bigger and bigger integrated circuits, but by making the individual components of the circuits smaller and smaller. The integrated circuits are constructed on thin wafers of silicon by a multistep process known as *photolithography*. Ultraviolet light (light with a shorter wavelength and higher frequency than visible light) is shone through a "mask" containing a pattern of transparent and opaque areas, and the resulting image is focused on a photosensitive film deposited earlier on the silicon surface. The chemical properties of that film are altered by light, and a chemical then selectively dissolves it, leaving a pattern of grooves matching the original pattern on the mask. That pattern of grooves is the first step to building the integrated circuit. Many more steps then follow, some optical, some chemical, some thermal (heating), some adding material, some removing material, until the desired three-dimensional arrangement of transistors and other circuit elements is formed. The silicon wafer may be up to a foot (30 cm) in diameter, and normally many identical circuits are built on its surface. The wafer is then sawed into many separate but identical pieces (chips), each now holding a complex integrated circuit to be inserted into your computer or other electronic device.

Over the half-century in which integrated circuits have been made by photolithography, the process has been steadily improved to produce smaller and smaller individual transistors and other components. The finest individual lines were originally many microns (μm, micrometers, millionths of a meter) in width, but the minimum feature width shrank gradually over the years to only a few microns, and eventually became less than a micron. Rather than being described in terms of fractions of microns, feature dimensions now are usually described in terms of *nanometers* (nm, billionths of a meter), a unit a thousand times smaller than a micron. When we introduced microns in Chapter 8, we asked you to think small. Now we ask you to think *very* small. Years ago, integrated circuits left the world of microtechnology and entered the world of *nanotechnology*. The minimum feature width in integrated circuits in production at time of writing has reached 45 nm (.045 μm), and it is still trending down.

What does this all have to do with magnetic levitation? To build those complex circuits on wafers of silicon, the wafers must be moved swiftly and extremely accurately, and smooth, friction-free motion is desirable. In addition to all the usual advantages of contact-free motion, here it also avoids the production, through wear, of fine particles that could wreak havoc if they found their way into the integrated circuit. In most modern lithography systems, the mask is moved during processing in the opposite direction from the motion of the wafers, and for motion of the mask, friction-free motion is also desirable. ASML, one of the current leaders in photolithography of integrated circuits, recently switched from mechanical and air bearings to magnetic bearings in their most advanced systems. For systems requiring processing in vacuum, clearly air bearings or other fluid bearings were inapplicable. For systems in air, the major advantage of mounting silicon wafers on maglev stages was improved control of forces and motions and a reduction in weight, which allowed greater accelerations and faster processing. Producing more chips per hour is a major selling point.

Details of the ASML maglev systems have not been fully released, but are probably similar to techniques developed and published earlier by researchers like David Trumper of MIT and Dan Galburt of Perkin-Elmer. In some designs, the stage carrying the silicon wafer is levitated and driven by forces between iron-free electromagnet coils on the levitated stage and neodymium permanent magnets on the stationary stage underneath. Although most electromagnets used in industry contain iron to increase their field strength, in this application the absence of iron both lightens the weight of the maglev stage and makes current-field relations linear. (These were also the advantages of using iron-free coils in the "flotors" in maglev haptics.) The neodymium magnets on the stationary stage below are arranged in what is known as a "Halbach array," a planar arrangement of a series of magnets magnetized in different directions that produces enhanced magnetic fields on one side of the array and near-zero fields on the other side. (The magnet array was named after its inventor, physicist Klaus Halbach.) The maglev stage is controlled in six degrees of freedom by currents to those coils (and the resulting magnetic fields and forces), and its position is monitored to great precision by feedback from sensors, probably both capacitive sensors and *laser interferometers.*

As noted earlier, capacitive sensors consist essentially of two charge-storing conducting plates separated by a thin insulating layer; the AC properties of the capacitor are extremely sensitive to the separation of the two plates, making it a good gap sensor. We have not encountered interferometers earlier, and these merit some explanation. In a laser interferometer, a laser beam on the stationary part of the stage is first split into two beams—call them beam A and beam B. Beam A travels a fixed path, while beam B is reflected off a mirror attached to the object whose position you want to sense, in this case the levitated wafer or mask stage. When beam B returns, it is remixed with beam A, and the "interference" between the beams A and B is detected. Light beams are waves of alternating electric and magnetic fields (we learned that from Maxwell in the nineteenth century), and before they were split, beams A and B were "in phase" with each other—that is, their electric and magnetic fields were increasing and decreasing in unison. But when they meet again, beam A and beam B have traveled different distances since they were split, and their electric and magnetic fields will no longer be alternating in unison. At some parts of the cycles of their alternating fields, the intensities of the two beams will add, while at other parts they will subtract. Adding the intensities of the two different beams will now give different overall results depending on how far beam B has traveled. That distance depends on the position of the mirror from which it was reflected, so the result of adding the intensities of the remixed beams A and B (i.e., their interference) will be a precise measure of the position of the mirror from which B was reflected (i.e., of the moving levitated stage).

Although capacitive sensors are very sensitive gap sensors, they have one conducting plate on the stationary stage and one on the moving platform, and therefore the range of platform motion for which they can be used is limited. A laser interferometer can work over large distances between the light splitter and the mirror, and so can operate over larger ranges of motion than capacitive sensors. Both types of sensors can determine position with extreme precision, to within less than a nanometer. Such precise position sensing is especially important for photolithographic processing of finely structured integrated circuits, since each processing step must overlay precisely the structures produced in the earlier steps.

While the maglev planar stages carrying silicon wafers during photolithography are indeed fully contact-free, floating on magnetic fields, their levitation heights are not several miles, the levitation heights of Gulliver's floating island of Laputa. They are not several inches like Martin Simon's magnet platform inside the pig's abdomen, not a centimeter or so like the magnet floating above a nitrogen-cooled high-temperature superconductor. They are often even less than a millimeter, the levitation height of the flake of diamagnetic graphite floating above a set of neodymium permanent magnets. The typical levitation heights of the maglev planar stages can be of the order of tenths of a millimeter, a few hundred microns. This is micro-levitation, perhaps not as exciting to the imagination as the floating island of Laputa. Professor Elena Lomonova of the Technical University of Eindhoven, the Netherlands, works on similar maglev planar stages and colorfully calls them flying carpets, but in terms of levitation height, such maglev stages are not nearly as impressive as the flying carpet of King Solomon or the one in Walt Disney's film *Aladdin*. However, they are very important commercially, doing successful nanoscale manufacturing in multimillion-dollar systems, and even at their modest levitation heights, they demonstrate the usefulness and wonder of magnetic levitation in fighting friction by fighting gravity.

Sub-Nano: Angstroms and Atoms

When I first introduced microns in Chapter 8, I asked you to think small. In the previous section, we introduced nanometers, which required thinking *very* small. Now I want you to think of the sizes of atoms, which are smaller than a nanometer. Think *very, very* small. Atomic sizes are often quoted in *picometers* (pm, trillionths of a meter), a unit a thousand times smaller than a nanometer. Atomic nuclei are much smaller than atoms, and their sizes are usually quoted in *femtometers* (fm, quadrillionths of a meter), units a thousand times smaller than a picometer and a million times smaller than a nanometer. But in this book, we don't have to think *that* small, since we are not concerned here with atomic nuclei. Since the diameters of atoms—the overall width of the volume filled by their electrons—are several hundred picometers, atomic sizes were traditionally quoted in a unit that is equivalent to 100 picometers or one-tenth of a nanometer—the *angstrom*. Atomic diameters vary from about 1 angstrom for helium (at the upper right of the periodic table of elements, home of the smallest atoms) to about 7 angstroms for cesium (at the lower left of the periodic table, home of the largest atoms). Thus the angstrom is a very useful unit.

The angstrom unit is not named after Rabbit Angstrom, protagonist of John Updike's popular novel series, but after Anders Jonas Ångström, a nineteenth-century Swedish physicist. He studied in detail the solar spectrum (i.e., the light emitted by the sun), and the unit that he and other scientists used to specify the wavelengths of the many components of sunlight was one-tenth of a nanometer. It became popularly known as the angstrom, the unit that in my college classes was used to express the size of atoms.

Gullible and trusting as I am, I had believed in atoms for some time without actually seeing them. But during my college years, the field ion microscope (FIM) was developed, the first experimental technique by which you could actually see individual atoms. The FIM revealed the arrangement of atoms only on the sharp tip of a limited number of metals in the presence of an extremely strong electric field and at low temperatures, but I recall being very impressed when I saw the first FIM images. I had been capable of believing in atoms without seeing them, since there was much experimental evidence for the existence of atoms before the FIM. But seeing images of atoms certainly helped. (Many years later, the doctoral thesis of one of my graduate students at MIT dealt with high-resolution electron microscopy, and the hundreds of images he produced revealed the detailed atomic structure of several metallic alloys we were studying. There has been lots of progress in science since my own college days a half-century ago.)

Several decades after the FIM, Gerd Binnig and Heinrich Rohrer developed the scanning tunneling microscope (STM), for which they received the Nobel Prize in Physics in 1986. Soon thereafter, the first atomic force microscope (AFM) was announced. The STM and AFM were able to reveal atomic structures on the surface of many more materials than the FIM, and without the need for extreme electric fields and low temperatures, they became much more generally useful. The basic physics of the STM and AFM differ. In both, a very sharp tip is brought into close proximity of a sample surface and sequentially scanned across the surface to produce a detailed image of the atomic structure of the surface, but the STM measures an electric current between the tip and the surface, while the AFM, as the name implies, measures a force exerted between the atoms of the tip and the atoms of the surface.

Both the STM and AFM can nicely resolve surface atomic structures. They have generated many fascinating images (including the famous image of the letters IBM spelled out in individual atoms—the smallest commercial advertisement yet), but in 1995, Dave Trumper of MIT and Bob Hocken of the University of North Carolina announced a maglev stage, designed by doctoral student

Michael Holmes, that could also *measure with precision* the sizes of various structures in integrated circuits, and in other products of nanotechnology, with resolutions on the atomic, and even sub-atomic, scale. They first called it the Angstrom Stage because it had angstrom resolution. Developed from earlier work with sensors, motors, and six-degree-of-freedom controls for maglev stages used for photolithography, it was the world's highest precision maglev stage (Figure 32). With further improvements, the Angstrom Stage evolved at North Carolina into the Sub-Atomic Measuring Machine (SAMM), with resolutions under some conditions as fine as 10 picometers (0.1 angstrom, .01 nanometer). Significantly sub-atomic.

I personally regret that the angstrom unit has gone out of favor. International standards organizations have officially discouraged its use and have instead encouraged replacing the term *angstrom* with

Figure 32. Sub-Atomic Measuring Machine, an extremely high-precision maglev stage. The upper block is levitated, moved, and controlled by four linear motors on the lower block.

its equivalent in standard metric terms, 0.1 nm or 100 pm. The relevant committees appear fixated on the series of prefixes for units smaller and smaller in factors of a thousand: milli, micro, nano, pico, and femto. But the angstrom as a unit, although it breaks that factors-of-thousand pattern (as does the centimeter), has the advantage of being of the order of the size of atoms—the basic building blocks of matter in our universe. I humbly suggest a compromise—to keep the angstrom alive by renaming it as the unit *angstrometer*. That is a four-syllable word ending in meter, which should satisfy the standards committees, and it keeps alive the memory of the angstrom, the unit in which I and others of my generation first learned about atoms. Another advantage: North Carolina's high-precision measuring machine, the SAMM, can be renamed the Sub-Angstrometer Measuring Machine without changing its acronym.

Maglev Rocket Sleds

We now return to thinking in terms of meters and kilometers instead of nanometers and angstrometers. In February 2008, a press release from Holloman Air Force Base in New Mexico announced, "test track takes levitation to new speeds . . . sending a rocket sled down the Maglev track with a speed of 423 mph (677 km/h), breaking the previous record of 361 mph (578 km/h) . . . held by Japan." That's something of an apples-to-oranges comparison, since the Japanese record was for an electromagnetically driven manned maglev train, while the Holloman record was for a small rocket-driven maglev sled. But they clarified that the goal of the maglev test was not to break speed records, but to reduce vibrations.

The 846th Test Squadron at Holloman AFB is the home of the world's leading hypersonic rocket sled track, nearly 10 miles long. The test track bridges the gap between laboratory testing and full-scale flight testing of advanced military weapon systems. One of the problems with earlier nonmaglev high-speed tests was that they produced unwanted vibrations. With the maglev sled, the fields from its superconducting electromagnets induced currents in copper plates in the track through relative motion, producing eddy-current damping that greatly reduced vibrations. Vibration forces are usually expressed in Gs, where one G is the force of gravity. For high-speed tests on their conventional track, they sometimes had vibration forces up to 80 Gs (more than any weapons were expected to experience in actual use), and the maglev test reduced vibration forces to no more than 6 Gs.

The maglev sleds held superconducting magnets built by General Atomics and wound with niobium–titanium wire originally intended for magnets in the infamous Superconducting SuperCollider (a monster underground particle accelerator started in Texas but eventually canceled by Congress in 1993 after about $2 billion had already been spent). Once the superconducting magnets were cooled to liquid helium temperatures and powered up to produce magnetic fields of about 7 tesla, leads were disconnected and the magnets put into persistent supercurrent mode before the rockets blasted the maglev sled off down the track, for a distance of about 130 yards, on its successful high-speed—and low-vibration—ride (Figure 33).

Although that maglev rocket sled went down the track pretty fast, rocket sleds at Holloman have traveled a lot faster. They have fired unmanned sleds up to speeds up to about 10,000 km/h (over 6,000 mph!) and in 1954 fired a rocket sled manned by John Stapp up to a speed of 1,017 km/h (631 mph), by far the record for a manned rail vehicle. The purpose of such manned tests was not to break speed records, but to see what forces of acceleration the human body could stand. Acceleration forces of 20 Gs, followed by deceleration forces of over 40 Gs as the sled braked, bloodied Stapp's eyes and produced lots of pain, but he survived. That earned Stapp, a surgeon and a Ph.D., the title of "fastest man on earth" and the cover of *Time* magazine. His body suffered no serious permanent damage from his record ride, and he lived to the ripe old age of 89. Among Stapp's later projects was promoting the use of seat belts in automobiles to limit the deceleration forces on the body in accidents.

Figure 33. Engineer inspecting maglev rocket stage after its successful travel down the test track at Holloman Air Force base.

Another project focused on propelling maglev vehicles down a track at high speeds was the Maglifter, an exploratory program at NASA's Marshall Space Flight Center in Huntsville, Alabama. The Maglifter was designed to study the possible use of electromagnetic propulsion and magnetic levitation to lower the cost of launching payloads into earth orbit. To launch payloads into orbit with normal rockets costs several thousand dollars per pound, partly because many, many pounds of fuel are required to launch each pound of payload. If the vehicle were first accelerated to very high speeds by magnetic forces, the amount of rocket fuel subsequently required to boost the vehicle into orbit would be much reduced, resulting in lowered launch costs. Under the direction of Maglifter champion John Mankins, two maglev tracks, each about 50 feet (15 m) long, were constructed at Marshall by two different companies. By the year 2000, each had demonstrated the rapid acceleration of a small maglev vehicle to a speed of about 60 mph (100 km/h) in a fraction of a second. As with the Holloman maglev rocket sleds, levitation was achieved with the electrodynamic approach— electromagnetic induction from relative motion. The vehicles held strong permanent magnets, and their motion down the track induced eddy currents in conductors in the track, resulting in upward repulsive forces and levitation. Propulsion was by linear electric motors of two different designs, with different arrangements of magnets and conductors, so magnetic forces both propelled and levitated the vehicles.

The Maglifter vehicles were not actually propelled by rockets, but were prototypes of vehicles that would later carry rockets. If the project had progressed further, it would have included electromagnetic propulsion of maglev vehicles to much higher velocities on a much longer track at Kennedy Space Center, and then firing of the rockets to launch the vehicles into orbit after high speeds had been reached. A third group funded in the Maglifter program was from Lawrence Livermore National Laboratory, and they proposed the use of Inductrack, a maglev system invented by Richard F. Post of Livermore. In the January 2000 issue of *Scientific American,* Post described Inductrack, which, like the ASML photolithography maglev stages, uses a Halbach array of neodymium permanent magnets to enhance the field from the magnets, although his magnets were much larger than those used in the photolithography stages. His article promotes the Inductrack approach for maglev trains, but also includes a diagram of a potential NASA launch-assist track and the claim that magnetic acceleration up to a speed of about 600 mph (1,000 km/h) before firing the rockets would reduce the needed rocket fuel by 30 to 40%. NASA budget limitations stalled the Maglifter program, but the idea of a maglev launch-assist system remains on the books at NASA, awaiting another champion and renewed funding. As one engineer who had worked on the Maglifter told me, the project is currently asleep, but not dead. And the general idea of electromagnetic launching is very much alive and remains under active study. The Navy will be introducing an electromagnetic aircraft launch system on its new aircraft carriers to replace steam catapults. The aircraft will be launched with a linear electric motor, as in the Maglifter system, but the Navy system is not maglev.

Although NASA's Maglifter program is currently on hold, and the Air Force's maglev rocket sleds at time of writing have not flown for over a year, these two programs featuring maglev vehicles flying down tracks at high speeds are a good lead-in to the topic of the next two chapters—maglev trains.

Flying Trains

Emile Bachelet

New York Times, March 15, 1912:

HE'D DRIVE CARS THROUGH AIR

Reporters were introduced yesterday afternoon to a puzzling, if not wonderful, new method of rapid transit—gravitationless, friction-less, electrical—the Bachelet method of electro-magnetic levitated transportation, which, if half of what its inventor expects of it is real-ized, will presently send whole carloads of passengers whizzing on invisible waves of electro-magnetism through space anywhere from 300 to 1,000 miles an hour.

The inventor who inspired the *Times* reporter to write these optimistic uplifting words nearly a century ago was Emile Bachelet, the first to patent and to demonstrate, via a model car magnetically levitated and propelled along a model track, a "puzzling, if not won-derful" *maglev train.* In the previous chapters, we have discussed a wide variety of examples of magnetic levitation, including floating globes and flying frogs, rotors and robots, wind tunnels and wind turbines. In this chapter and the next, we turn to the application most commonly associated with the term *maglev* in the minds of the general public—maglev trains.

Trains were first added to the transportation options for people and goods in the early nineteenth century. Passenger trains were introduced in the United States in 1830, and by 1869 the intercontinental railway was completed. In 1902, a decade before Bachelet's demonstration in New York, the New York Central introduced its famous 20th Century Limited, an express passenger train service that traveled the 960 miles between New York and Chicago in 20 hours, averaging (including station stops) 48 miles per hour (77 km/h). In later years, the travel time was shortened to 16 hours, with an average speed of 61 mph. Like all other trains before it, and like the recently introduced automobile, the 20th Century Limited depended on one of early humankind's most important inventions—the wheel. Bachelet's remarkable proposal was to do away with our long-standing dependence on the wheel, and transport people and goods through the air at high speeds, floating on magnetic fields.

Although the specific method of "electro-magnetic levitated transportation" that he developed never took off commercially, Emile Bachelet can nevertheless reasonably be considered the father of the maglev train. He was born in France in 1863, emigrated to the United States in about 1881, and became an electrician, first in Boston for MIT (long before its move to Cambridge) and later for the city of Tacoma, Washington. There Bachelet reportedly noticed that his arthritic pains went away when he was near large electrical generators, and he began to experiment with therapeutic uses of electromagnets. He obtained several patents for electromagnetic medical devices and received financial support that allowed him to move to New York City in about 1905 and set up the Bachelet Medical Apparatus Company, with a factory in Brooklyn and an office and laboratory in Manhattan. In experiments in his lab, he became more and more fascinated with electromagnetic forces, and developed demonstrations for vaudeville shows he called "The Bachelet Mystery." One of his "mysteries" included a clay model of a hand activated by hidden electromagnets that could be energized offstage to move the entire hand or just its fingers in response to questions from the audience, and even to levitate the hand. These magical novelties provided some income, but he became most interested in his invention, patent granted in 1912, of a "levitating transmitting apparatus" that soon became known to the public as his "Flying Train."

At his demonstration in New York in March 1912, Bachelet first demonstrated the basic physical principles of his maglev train by levitating an aluminum plate above an AC electromagnet. The alternating magnetic fields from the electromagnet generated alternating eddy currents in the plate and repulsive forces like those that lifted the aluminum frying pan in Figure 13. Something of a showman from his experiences on the vaudeville circuit, Bachelet then placed a small glass bowl, with several live goldfish swimming in it, over the electromagnetic coil. The bowl also held a small aluminum plate sitting on the bottom, and when Bachelet turned on the current to the AC electromagnet, the aluminum plate rose up through the water, the goldfish dodging around and under it as it rose, and the plate then hovered motionless above the water.

Having thus impressed his audience with the magic of magnetic levitation, Bachelet proceeded to the main demonstration, his maglev car. It was an aluminum cylinder about 3 feet (nearly a meter) long, with tapered ends, that he levitated and moved with magnetic forces along a track about 35 feet (11 m) long. The physics of his maglev model train was a bit more complex than in his demonstrations with the aluminum plates, since magnetic forces from AC electromagnets in the track acting on the car produced both levitation (maglev) and propulsion (magprop?). Bachelet noted that other means of propulsion could be used, but since the car was levitating, traditional means of propelling trains by turning wheels in contact with track were clearly unavailable. Most electric motors spin a rotor with rotational magnetic forces (torques) exerted between the rotor and the stator. Bachelet's propulsion system was a form of *linear motor,* which can be viewed as produced by cutting open the circular stator of an ordinary motor and spreading it out along the track to produce linear rather than rotational motion. More about linear motors will be presented later in this chapter.

Bachelet was seeking investors to support further development of his maglev train, and John Jacob Astor was among those who became interested. Astor visited Bachelet at his home and reportedly promised to fund a full-scale laboratory after he returned from a brief trip to Europe. But Astor booked his return trip on the *Titanic,* so Bachelet's likely source of funding sank in the Atlantic Ocean that April. A British investor encouraged Bachelet to come to London, originally to develop his miniature railway into a mail delivery system. But Bachelet had bigger plans. He appears in the *Times* archives again in May 1914, with a brief report of his demonstration in London of his "model air railway" to officials of the British Admiralty. Among them was 40-year-old Winston Churchill, and he can be seen admiring Bachelet's maglev system in Figure 34.

Figure 34. Officials of the British Admiralty in 1914 inspecting the model maglev train of Emile Bachelet. Winston Churchill, then 40, can be seen at the center (to the left of the blur).

"By George, it's great!" said Churchill, "It is the most wonderful thing I have ever seen." A more practical engineer in the group said that the "model was interesting, but it was a long way from a model to a full-sized railway." In early June, a prospectus offering shares in Bachelet's "levitation railway project" was issued, but later that month World War I broke out in Europe and drew attention away from flying trains.

A 1917 issue of *The Electrical Experimenter*, a U.S. magazine with the "largest circulation of any electrical publication," featured a cover announcing optimistically, "Electric-Flyer makes 500 miles an hour" (Figure 35). Recognizing that with Bachelet's system the track "would cost a small fortune to build," the editors suggested that a more practical approach would be to put the electromagnets in the car and the aluminum conductors on the track, the reverse of Bachelet's arrangement. By 1921, the few who had invested money in Bachelet's project had lost their enthusiasm, funding dried up, and his maglev train never progressed to a full-scale railway. Its major technical limitation was that it needed a large amount of electrical power even for a mail delivery system, and would require an inordinate amount of power for full-scale freight and passenger transportation. Bachelet stubbornly believed he could in some way harness "free energy" to solve this problem, but never found the way, and died in 1946 at the age of 83.

In his 1987 book on the history of linear electric motors, Eric Laithwaite wrote that Bachelet "was too far ahead of his time to be successful" but that it was he who "first conceived the idea of an object in free flight—levitated, guided and propelled by the mystic forces of electromagnetism which Faraday had revealed almost a century earlier." Laithwaite was a British pioneer of linear electric motors and magnetic levitation. Had he been an American, he might have credited Joseph Henry with sharing with Faraday the discovery of the "mystic forces of electromagnetism."

Robert H. Goddard, American pioneer in rocketry, apparently first learned about Bachelet's "flying train" in 1914 by reading about

Figure 35. Cover of March 1917 issue of *Electrical Experimenter,* which includes an editorial suggesting a design variation for Bachelet's maglev train.

it in the *Boston Globe*. In response, Goddard submitted an article to the *Journal* of Worcester Polytechnic Institute, his alma mater, entitled "Bachelet's Frictionless Railway at Basis a Tech Idea." In it, Goddard states that he himself first conceived of a high-speed magnetically levitated train in 1904, when he wrote an essay for his freshman English class on the topic of future means of travel. Then only 22, his already inventive mind conceived of a high-speed train both levitated and propelled by the "magic power of magnetism." But he went one step further than Bachelet. "Although Bachelet's claim, may, at first sight, seem extravagant," wrote Goddard, "it actually does not give the method credit for the speed it is capable of delivering." Goddard's freshman essay, published in condensed form in the *Scientific American* in 1909, not only proposed magnetic levitation to eliminate friction of wheel on track, it also proposed having the maglev train travel in an evacuated tube to eliminate air resistance, which becomes a limiting factor at very high speeds. In the WPI *Journal* article, he also reproduced a short science-fiction story he had written in 1906, "The High-Speed Bet," that featured a 1958 trip on such a train from Boston to New York, covering the 200 miles in just 10 minutes, for an average speed of 1,200 mph (1,930 km/h). Since he assumed constant acceleration for the first half of the trip and constant deceleration for the second half, the maximum speed the train reached at the midpoint was double the average, or 2,400 mph. An imaginative essay and a short story written during Goddard's undergraduate years at WPI do not challenge Bachelet's achievement of patenting, building, and demonstrating the first serious model of a maglev train. However, Goddard did later achieve remarkable success in rocketry, so perhaps if he had instead maintained interest in developing maglev trains, the United States might be much farther along than it is today.

Kemper and Electromagnetic Levitation (EML)

A second engineer with a reasonable claim to the title of father of the maglev train is Hermann Kemper of Germany. Although his experiments and demonstrations were later than Bachelet's, his technical approach—levitation by feedback-controlled attractive forces—provided the basis for the subsequent development of the Transrapid, the first high-speed maglev train to reach commercial use. The basic principle of Kemper's magnetic levitation was that of the floating globe desk toys (Figure 20), an upward force of attraction that achieves vertical stability with a sensor and feedback circuit that controls the current to an electromagnet. This general approach to maglev is often identified with the acronym EMS for electromagnetic *suspension*, but for this book on maglev, I prefer EML for electromagnetic *levitation*.

Kemper was born in 1892 in Nortrup, a small town in northern Germany, and reportedly started thinking about magnetic levitation as early as 1922, while still a student in Hannover. Perhaps he had heard of Bachelet's demonstrations in London a few years earlier. In 1927, he took over his family meat products company in Nortrup, which provided him the financial means to explore his ideas experimentally, and in 1934 he received a patent for a "suspension train without wheels, propelled along iron tracks by a magnetic field." In the following year, Kemper built and demonstrated a prototype system, levitating 210 kilograms (about 100 lb) with a power consumption of only 270 watts. For the feedback circuit, he used a capacitive sensor to measure the air gap. Because this was well before the invention of transistors, his electronic control circuits relied on vacuum tubes. He followed this demonstration with promotion of his results and ideas in a 1938 paper, arguing that magnetic levitation was a "fundamental new means of locomotion" (*grundsätzlich neue Fortbewegungsart*). The technical experts in Hitler's Germany apparently found other innovations of more interest than Kemper's maglev dreams, but Kemper persisted. By the 1960s, industry in West Germany had largely recovered from the devastation of World War II, thanks in large part to the Marshall Plan, and Kemper's ideas began to take hold there and elsewhere.

The first full-scale maglev system capable of holding passengers was the Magnetmobil built by Messerschmitt-Bölkow-Blohm (MBB) and demonstrated in 1971. The vehicle weighed about 6 tons, could carry up to ten passengers, and reached a maximum operating speed of about 100 km/h (60 mph). Another German company, Krauss-Maffei (KM), built a maglev system with a somewhat different levitation and propulsion system, which led to the first Transrapid test vehicle, TR-01, in 1969. That was developed through several stages until the TR-04, an 18.5-ton vehicle that in 1973 reached 253 km/h (152 mph) on a 2.4-km (1.4-mile) track. That year, MBB and KM merged to form the Transrapid EMS Consortium, which several other companies soon joined. Corporate participation in the project varied over the years, with Siemens and Thyssen Krupp later becoming the major players.

(I'm already getting tired of quoting distances in both kilometers and miles, and speeds in both km/h and mph, and perhaps you are getting tired of that too. I'll be referring to both distances and speeds a lot in this chapter and the next, and perhaps it will be best if I simply remind you that a kilometer is about six-tenths of a mile: 1 km = 0.621 mile and 1 mile = 1.61 km. When I'm driving in Europe and need to translate one way or the other, I use the rough factors 0.6 and 1.6, which are good enough for most purposes. So in the rest of this chapter and the next, I'll use miles and mph for distances and speeds in the United States and kilometers and km/h elsewhere, and trust that you can apply the rough factors 1.6 or 0.6 to translate when you feel the need.)

In 1972, the German government awarded Kemper a prestigious national award for service to Germany in his half-century of effort toward developing maglev transportation. Five years later, they made the decision to cease funding of competitive maglev technologies and to concentrate further effort exclusively on the EML system pioneered by Kemper. As fate would have it, that was the year of his death. But unlike Emile Bachelet, Hermann Kemper was fortunate to have lived long enough to see his early maglev dreams become real in the form of multi-ton high-speed maglev trains, albeit demonstration models.

The detailed designs of Transrapid trains continued to evolve, but levitation was still based on servo-controlled upward attractive forces. The bottoms of the cars wrap around the edges of the T-shaped track and contain electromagnets that are attracted upward toward the track (Figure 36). Sensors and feedback circuits maintain a gap between the electromagnets and the track above them of about 1 cm, and similarly maintain a sideways gap to provide guidance and stability in the lateral direction. Powered AC electromagnets in the track produce

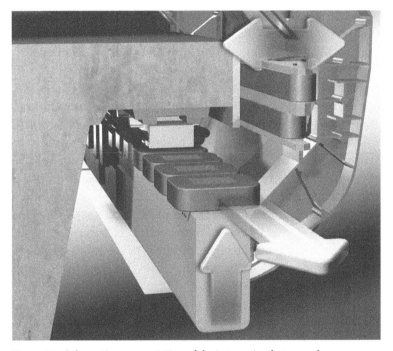

Figure 36. Schematic representation of the interaction between the Transrapid track (left) and the magnets in the wraparound portion of the car (right). Represented by three arrows, vertical forces produce controlled levitation, lateral forces produce guidance, and forces along the track produce propulsion (linear synchronous motor). Photo courtesy of Transrapid International GmbH & Co. KG, © Fritz Stoiber, photographer.

forces on the DC electromagnets on the car to move it along the track in a propulsion system known as a *linear synchronous motor*. Current changes in the track change the polarity of track electromagnets in synchronism with the motion of the car, pushing and pulling it forward on a traveling wave of magnetic fields. (A linear synchronous motor also drives many amusement park rides, like the Superman ride of Six Flags Magic Mountain, but here the magnets on the cars are permanent magnets and no maglev is involved.)

There are numerous forms of rotary and linear electric motors. A standard rotary *induction motor* uses a rotating magnetic field produced by AC electromagnets on the stator to induce currents in a conducting rotor and the resulting torque (the rotational force exerted on the induced currents by the rotating field) to spin the rotor. It is sometimes called a rotating transformer, with the stator the primary and the rotor the secondary. In a standard *synchronous motor,* the rotor is instead a magnet that rotates in synchronism with the rotating field from the stator. Maglev trains are propelled either by a linear induction motor or by a linear synchronous motor. In synchronous systems, usually the magnets on the car are permanent magnets or DC electromagnets and the powered AC electromagnets that produce the propulsion are in the track. When induction motors are used for propulsion, the powered AC electromagnets (equivalent to the stator in a standard rotary induction motor) are usually in the car and the conductors (equivalent to the rotor) are in the track, but the reverse is also possible, as in the propulsion system used by Bachelet.

Kemper lived long enough to see the TR04 in action, but not the TR05, the first maglev train to receive full certification for carrying passengers, which in 1979 smoothly carried many thousands of people in short low-speed (75 km/h) trips on an elevated railway between the parking lot and exhibition hall at the International Transportation Fair in Hamburg. Transrapid next built an elevated maglev test track in farmland of the Emsland district of West Germany, not far from the Dutch border. The track had a total length of

31.5 kilometers, including two loops, two switches, and a straight run of 24 kilometers between the two switches. There they tested the TR06, a two-car, hundred-ton vehicle that had test runs of steadily increasing speeds until they reached 413 km/h in 1988, then a world record for a manned vehicle. (The total weight levitated also handily beats the 35 tons lifted by the S2M magnetic bearing for the Japanese water turbine discussed in Chapter 8.) In 1993, the TR07 set a new speed record of 450 km/h there, and also ran a nonstop endurance run of over a thousand miles. Two years later, they began offering exciting high-speed rides to visitors, which became so popular they had to expand service to eight rides a day, six days a week. Films of the Emsland test track show cows in a nearby field grazing undisturbed as a train speeds past, not so subtly making the point that the sound levels generated by high-speed maglev trains are not extreme. The sounds generated by maglev trains are generally lower than those generated by wheel-on-rail trains traveling at the same speed, but that difference is small at very high speeds, when most of the sound results from air flow around and under the vehicle.

In addition to Transrapid, the EML approach to maglev trains was explored by other groups in the 1970s and later, sometimes only to the design stage or the stage of small demonstration vehicles. These included several universities, Rohr Industries of California (whose ROMAG technology was later transferred to Boeing), Ford, General Motors, and Grumman, whose design used a superconducting magnet to achieve a wider gap between the vehicle and the track. British Rail's EML project led to a low-speed maglev vehicle that operated from 1984 to 1995 at the Birmingham Airport, linking the airport terminal with the rail station. But the most extensive EML effort outside of Germany was the HSST (High-Speed Surface Transportation) program initiated in 1972 by Japan Airlines. Despite its original name, the HSST program eventually evolved into the low-speed Linimo system that is currently operating commercially in Nagoya as "urban maglev," a topic we consider in the next chapter.

Powell/Danby and Electrodynamic Levitation (EDL)

According to James Powell, then a scientist at Brookhaven National Laboratory, Long Island, New York, he first started thinking about maglev trains in 1961, while he was stuck in a multi-hour traffic jam at the Throgs Neck Bridge, the easternmost of the many bridges on which motorists enter or exit the island. "It took four hours to get over the bridge, a total distance of three miles, "said Powell, "It was then that I thought of maglev." As it happened, 1961 was also the year that high-field superconductors suddenly became a hot topic, and Bell Labs, Westinghouse, GE, and other groups were racing to produce the first high-field superconducting electromagnets, which would soon be capable of producing much stronger magnetic fields than those possible with permanent magnets or conventional iron-core copper-wound electromagnets. After Powell's frustrating experience with traffic congestion, he and his Brookhaven colleague Gordon Danby began brainstorming about ways to magnetically levitate a train, thereby achieving the possibility of achieving high speeds without the friction of wheel on rail. They were familiar with the fact, discussed in Chapter 5, that a magnet moving relative to an electrical conductor induces eddy currents and repulsive forces. Perhaps if a train were carrying high-field superconducting electromagnets and moving fast enough, it could induce eddy currents in an underlying conducting track strong enough to produce levitation of even a multi-ton vehicle. In 1967, Powell and Danby published "A 300-MPH Magnetically Suspended Train" and filed for a patent detailing their ideas, which was granted in 1969. Among the prior patents they cited in their application was the 1912 patent by Bachelet. The Powell/Danby system of levitation operates only when the train is moving rapidly and is therefore known as electro*dynamic* levitation or EDL, in contrast to the EML attractive feedback method originated by Kemper, which can levitate even when the train is standing still. (Others prefer the acronyms EMS and EDS. We could compromise and just call the competing systems EM and ED, but the latter term unfortunately suggests insufficient lift.)

The most extensive project employing EDL is that initiated in the 1960s by Japan National Railway. By 1977, they had built a 7-kilometer test track on the southern island of Kyushu, and two years later, their unmanned 10-ton ML-500 achieved there a world speed record of 517 km/h. In the following decades, the Japanese EDL trains and the German EML trains competed for world speed records for unmanned and manned vehicles, with EDL eventually taking and keeping the lead, particularly after completing the longer (18.4 km) Yamanashi test track in a mountainous region on the main island of Honshu. This track may some day become part of a high-speed maglev line linking Tokyo and Osaka. In 2003, they achieved here a speed of 581 km/h in a manned three-car train, a speed formally recognized by Guinness as the current world record for manned trains. The Yamanashi test track has two guideways, allowing testing of two high-speed trains passing in opposite directions. In 2004, two trains passed each other with a relative passing speed of 1,026 km/h—with no ill effect.

The superconducting electromagnets in the cars of the Japanese EDL trains are wound with low-temperature high-field superconducting wire (an alloy of niobium and titanium), are cooled with liquid helium, and create a magnetic field of over 4 tesla. (Later discovery of high-temperature superconductors encouraged some to hope that they could be used for EDL trains without the need for liquid helium, but high-temperature superconductors have not yet been found capable of producing nitrogen-cooled electromagnets competitive with helium-cooled electromagnets wound with niobium–titanium.) The EDL car necessarily starts on wheels, which lift when the relative speed between car and track is sufficient to produce levitation through currents induced in the track. I have not yet had the pleasure of a demonstration ride on the Japanese EDL train, but a colleague tells me that it is exciting when the train reaches sufficient speed to "take off," even though the train elevates only a few inches. Electromagnetic interaction between the superconducting magnets in the cars and nonsuperconducting coils in the U-shaped track produces both levitation and lateral guidance, and results in propulsion via a linear synchronous motor. The gap between train and track in the EDL system depends on speed and can reach several inches, far greater than the roughly half-inch gap in the EML Transrapid and its Japanese counterpart (HSST). This difference in train-track gap is said to be one of the factors that led Japan to choose the EDL system for its high-speed trains on the basis of safety, since Japan is earthquake country.

For several years in the 1970s, German industry, supported by government funding, also explored EDL. In Erlangen, Bavaria, they built a circular test track 280 meters in diameter, banked at an angle of 45 degrees to allow continuous operation at high speeds. They first tested propulsion systems with wheels supporting the 18-ton vehicles, and later installed eight helium-cooled superconducting magnets for levitation (to a 10-cm, 4-in. gap) and guidance (lateral stability). Although they successfully achieved design speeds of 200 km/h, Germany in 1977 decided to drop EDL and focus its further development of high-speed maglev trains on the EML Transrapid.

In the United States, the major EDL project in the early 1970s was the Magneplane, developed by Henry Kolm and Richard Thornton at MIT's National Magnet Laboratory with modest industrial and government support. The cylinder-shaped Magneplane was designed to fly about a foot above a semicircular track that allowed the vehicle to self-bank on curves. In a 1973 article in *Scientific American* entitled "Electromagnetic Flight," Kolm and Thornton promoted maglev trains in general and their Magneplane in particular. They built a 1/25-scale vehicle and in 1974 flew it down a similarly scaled linear test track at 60 mph. Propelled by linear synchronous motor, the Magneplane carried superconducting niobium–tin electromagnets that produced levitation via repulsion from eddy currents induced in aluminum in the track. However, U.S. maglev funding dried up, and plans for a full-scale model of the Magneplane did not materialize.

Throughout the 1980s, development and testing of full-scale maglev trains continued in Germany and Japan, but maglev languished in the United States until the 1990 formation of the National Maglev Initiative, a cooperative effort of the U.S. Departments of Transportation and Energy and the U.S. Army Corps of Engineers. Maglev enthusiasts had generated an effective champion in Senator Daniel Patrick Moynihan of New York, who wrote an op-ed piece in the November 1989 issue of *Scientific American* arguing that maglev "will define the coming century much as the railroad defined the last one and the automobile and airplane have defined this one." And $751 million of a 1991 $151 billion transportation act was authorized for maglev, pending the initial results of the more modestly funded National Maglev Initiative, but Congress never carried through on that level of funding. Four companies were awarded research and development grants, all based on achieving maglev with the use of superconducting magnets, clearly under the influence of Powell and Danby. As noted earlier, Grumman's team designed an attractive feedback-controlled EML system, but the three others— Magneplane, Foster-Miller, and Bechtel—proposed repulsive EDL systems of varying designs. The final report of the National Maglev Initiative, published in 1993, included details of the reports of the four companies and, after considering sharing of German or Japanese technologies, concluded optimistically that U.S. industry was fully capable of developing advanced maglev technology on its own. It concluded further that several U.S. transportation corridors had the potential for sufficient revenues to cover operating costs and a substantial portion of the initial capital costs. But there's always a catch—in this case, an important one. The report also concluded that a U.S. maglev system was "not likely to be developed without significant Federal Government investment." The funding for the four initial studies under the National Maglev Initiative had been $36 million, but the next step would require much more. Senator Moynihan would have to wait for the transportation innovation that he felt "will define the coming century."

In his 1991 book *Supertrains*, Joseph Vranich wrote confidently, "the first super-speed maglev line in America will be built in Florida by Maglev Transit Incorporated, a German-Japanese-American consortium." A 1992 *Scientific American* article similarly predicted, "the first commercial line is likely to be built in the U.S., which is seen as perhaps the primary market for the Transrapid." The plan was for a 14-mile airport connector at Orlando, and originally it was to use Transrapid technology and reach a maximum speed of 250 mph. Scheduled to open in the fall of 1995, it was to be privately funded, and the president of Maglev Transit claimed that the project was "financially feasible without any public subsidy." Orlando, home of Disney World's Magic Kingdom and Tomorrowland, seemed like an ideal place for a futuristic transportation system like maglev. But problems arose, and by 1993, plans had shifted from Transrapid technology to Japanese HSST technology and a maximum speed of 130 mph. Financial problems and business disagreements deepened, and in 1994, the local paper finally reported, "Orlando Maglev Dead in Its Tracks." Florida was also the home of another industrial hopeful, American Maglev Technology, Inc., a company that built a small maglev test track in Edgewater with public subsidy in the form of grants from federal, state, and county governments. But they also ran into financial problems, and by 1996, newspaper accounts about them reported, "Maglev Off Track as Money Runs Out." The century ended with U.S. maglev still way behind the Germans and Japanese. The name of the famed New York Central train mentioned earlier provides a concise report on the century's progress of U.S. maglev trains: 20th Century—Limited. Very Limited indeed.

The Competition

Despite relative inaction in the United States, government and industry in both Germany and Japan in the twentieth century cooperated to invest billions in the development of maglev trains. The technology choice for high-speed maglev trains narrowed down to electromagnetic vs. electrodynamic (EML vs. EDL)—that is, Kemper's feedback-controlled magnetic attraction, favored by Germany, vs. Powell and Danby's motion-induced magnetic repulsion, favored by Japan (Figure 37). The many steps of maglev development had also reached numerous other decisions along the way, for example, propulsion by induction motors vs. synchronous motors (the latter favored by both Germany and Japan for their high-speed trains), detailed designs of guideways and vehicles, vibration control and ride comfort, vehicle behavior on grades and curves, lateral guidance, electronic control systems, switching, and so on. But by the end of the century, despite considerable technical progress, and although thousands of people had experienced high-speed maglev demonstration rides in each country, neither Germany nor Japan had progressed to maglev commercial passenger service. In both countries, as throughout the world, maglev trains were also competing with many other demands for private and public funding.

With regard to overall spending for transportation, the major competitors included the two transportation options that Senator Moynihan noted had defined the twentieth century—automobiles and airplanes. I personally love the U.S. Interstate Highway System initiated by President Eisenhower, but it took 35 years to complete and, adjusted for inflation, cost nearly half a trillion dollars in 2008 dollars. (Some estimates are even higher.) The majority of government spending on transportation comes from state and local governments, and for the year 2000, the total transportation expenses by all levels of government in the United States were about $130 billion on highways, $21 billion on airports, and $1 billion on rail. We are a car-centered culture. But highway congestion has been a major problem for many years, as James Powell memorably learned in 1961, and recent experience shows that highway improvements increase highway usage but seldom significantly reduce congestion.

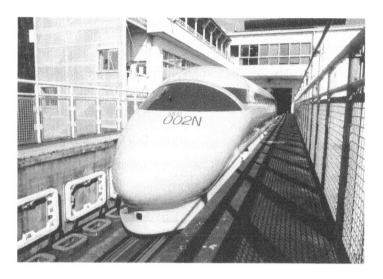

Figure 37. Two full-scale maglev trains that carried many passengers on high-speed demonstration rides on test tracks in the twentieth century. The Japanese MLU002 (top), an EDL system, contains superconducting electromagnets and is levitated by repulsive forces produced by current induced in the guideway coils by the moving train. Germany's Transrapid (bottom), an EML system, is levitated by attractive forces between iron-core electromagnets on the cars and the steel track.

Government studies that compare the competitiveness of autos, planes, and trains for personal travel suggest that even moderate-speed trains compete well for travel distances between about 100 miles and 600 miles. From my current home in a Boston suburb, I drive my car to nearby Cape Cod and fly on planes to faraway Florida or California. But I live roughly 200 miles from New York City, and for that intermediate-length trip I have several choices. I have driven there and I have flown there, but I much prefer the train for convenience, comfort, and cost. Thinking more broadly, our urgent current needs to reduce our dependence on foreign oil and to slow climate change by reducing emission of greenhouse gases both argue strongly for train travel to replace gas-guzzling planes and cars. Compared to planes and cars, trains are "green." Trains also win on the weather front; when your flights are canceled, the trains usually remain running. For safety, the train beats the car for lower deaths per passenger mile. And the faster the train can travel, the better it looks in the competition against both planes and cars.

However, in the overall competition for government financial support of transportation, one of the most threatening competitors to high-speed maglev trains is *other trains,* that is, high-speed wheel-on-rail trains. Since the 1960s, when maglev trains first began to be seriously developed, there has been considerable improvement in wheel-on-rail systems. As noted in our Preface, technology is a moving target.

The first of the modern high-speed rail lines was Japan's Shinkansen, the so-called bullet train that started to run between Tokyo and Osaka in 1964, just in time for the Tokyo Olympics. Maximum operating speed then was 220 km/h and today is 270 km/h. In a 1996 test run, these trains reached a speed of 443 km/h. Despite carrying up to 150 million passengers a year on the Tokyo-Osaka line, the Shinkansen has a nearly perfect record for punctuality and a remarkable record for safety, with no deaths in over four decades of service (other than suicides jumping from or in front of trains). Over the years, Shinkansen expanded to a rail network that links most major cities on the islands of Honshu and Kyushu. Some parts of the network have operating speeds up to 300 km/h. Most Shinkansen lines run on dedicated tracks that are not used by freight trains or slower local trains, have no grade crossings, and have only very gentle curves.

The fastest wheel-on-rail line today is the French TGV (Train á Grande Vitesse, i.e., "high-speed train"). High-speed passenger service between Paris and Lyon has been offered since 1981, and other TGV lines have since been added to build a wide network. In 2007, a special TGV test train set the world speed record for a wheeled train, reaching 575 km/h. That really *is* a Grande Vitesse—just 6 km/h below the world speed record for the Japanese EDL maglev train on the Yamanashi line! TGV also holds the record for the fastest commercially scheduled train trip of 279 km/h. Germany's high-speed wheeled train is the Intercity-Express (ICE), which has operating speeds comparable to the TGV. I personally have had the pleasure of riding on the Shinkansen, TGV, and ICE, and the rides are generally smooth, comfortable, and reliable as well as fast. That's the way to run a railroad.

In the United States, which in many other areas is still the world's technology leader, our Federal Railroad Administration (FRA) claims that we do have *one* passenger-carrying "high-speed train"—Amtrak's Acela Express running in the "Northeast Corridor" between Boston and Washington, D.C. The Acela, which began operation in 2000, is considered a high-speed line by the FRA because it is capable of speeds up to 150 mph (241 km/h) on portions of its track. When I first started using the Acela for my occasional trips between Boston and New York, a voice would come on the public-address system a little north of Providence announcing that we were now traveling at 150 miles per hour. However, on portions of Acela's travel through Connecticut, I could look out the window and see cars passing us on the adjacent Connecticut Turnpike. The average speed of the Acela on its Boston–New York run, between delays for stops, is only about 80 mph. My mathematics background tells me that since it goes faster than that north of Providence, it must run slower than that in other places, like where those cars were passing us in Connecticut. However the FRA may categorize the Acela, for most of my ride it did not feel much like a "high-speed train." Much of the reason is the track, which has numerous grade crossings and sharp curves. (To help handle the curves, Acela cars are built to tilt.) Whereas the Shinkansen, TGV, and most other high-speed lines in the world were built on entirely new railways, dedicated to carrying only high-speed trains, the Acela was not. Acela cars are also heavier than the cars of high-speed trains in other countries, partly to meet various stringent FRA regulations.

Although I do sincerely wish that the Acela traveled faster than the cars I saw passing us on the Connecticut Turnpike, for balance I should note that I did not actually *envy* the drivers of those cars. I have driven in heavy traffic on the Connecticut Turnpike myself, and for me, riding the Acela while reading a good book, and occasionally closing my eyes when I want a brief break from reading, is a much more pleasant experience. But how about maglev, hailed by Senator Moynihan in 1989 as the transportation system that "will define the coming century much as the railroad defined the last one and the automobile and airplane have defined this one"? The next chapter considers maglev progress, or lack of it, in the early years of the twenty-first century.

All Aboard!

Urban Maglev

From their earliest days, maglev trains have been associated with the vision of *high-speed* transport. After seeing Bachelet's 1912 demonstration of his model maglev train, the *Times* reporter speculated that it might "presently send whole carloads of passengers whizzing on invisible waves of electro-magnetism through space anywhere from 300 to 1,000 miles an hour." In the previous chapter, my discussions of German and Japanese progress with maglev trains frequently referred to their achievements of higher and higher speed records. But many people believe that maglev trains also have promise for *low-speed* transport in an arena that has become known as "urban maglev." Here proponents quote maximum speeds of only up to 100 mph, with average operating speeds considerably less. Urban transport typically involves numerous stops, and the arguments for maglev here focus not on high speed but on smooth, quiet, safe, reliable, pollution-free, energy-efficient, and cost-effective (low maintenance and operating costs) transport of many people per hour. The ability to cope with grades and tight curves is also an important concern.

Whereas high-speed maglev has been an on-again, off-again interest of the Federal Railroad Administration (FRA), study of urban maglev falls under another division of the U.S. Department of Transportation, the Federal Transit Administration (FTA). In January 1999, the FTA announced its urban maglev program, which led to the selection and funding of five U.S. groups that convened in Washington in 2005 to report the results of their projects. An important foreign invitee to that meeting was a representative of Japan's Linimo urban maglev line, which has been carrying many millions of passengers on the outskirts of Nagoya since 2005. As discussed in the previous chapter, Linimo, which claims to be the world's first commercial urban maglev, is an outgrowth of the Japanese HSST program. It uses the EML feedback-control system with U-shaped iron-core electromagnets providing the upward attractive force, and is propelled by a linear induction motor. It currently has nine stations along its 8.9 km line, which starts from a terminal of the Nagoya subway line and travels about 1.6 km through a tunnel, and the remainder of the line on an elevated guideway, to the former site of the Expo 2005 fair. The three-car trains travel at a maximum speed of 105 km/h and carried over 20 million passengers in 2005 during the Expo. At the 2005 FTA meeting in Washington, a representative from Korea also reported on a technologically similar but shorter low-speed maglev system operating in South Korea near the city of Daejeon.

Of the five U.S. teams that received FTA funding in their urban maglev program, two primarily studied the possibility of using Linimo HSST technology in Maryland or in Colorado, while the other three focused mostly on hardware development. One of these was Maglev2000, a Florida firm founded by Robert Powell and Gordon Danby, who in the 1960s had first introduced the idea of using superconducting electromagnets and the EDL system—levitation by motion-induced eddy-current repulsion—for high-speed maglev trains. For that, Powell and Danby were awarded the prestigious Franklin Medal by Philadelphia's Franklin Institute in 2000. (As noted on their website, "previous Franklin Medal awardees include Nikolai Tesla, Charles Steinmetz, and Albert Einstein." Impressive company indeed!) Although Maglev2000 originally was focused on high-speed maglev, they shifted to low-speed maglev to receive FTA funding. But according to the final 2009 FTA report on "lessons learned" from their initial urban maglev program, Maglev2000's technical efforts to achieve levitation of their prototype model with superconducting electromagnets led to "one failure after another." The Powell/Danby 1960s concept of using helium-cooled super-conducting electromagnets to achieve levitation remains alive in Japan but has lost ground elsewhere.

Two other teams achieved more technical success. General Atomics chose to use the Inductrack system, originated by Richard Post of Lawrence Livermore Lab, which employs the EDL approach with neodymium permanent magnets in Halbach arrays rather than with superconducting electromagnets. Magnet arrays on the vehicle above and below conducting coils in the track provide levitation by induced currents through relative motion, and another set of magnets on the vehicle is used for propulsion and guidance. With their own and FTA funding, General Atomics built a 130-yard test track at their site in San Diego, California, and a full-scale lightweight vehicle that operates with a levitation gap of roughly an inch once it reached modest speeds. They made an environmental assessment of the possible deployment of their system on a proposed track on the campus of California University of Pennsylvania, and are also pro-moting the use of their urban maglev system for short-haul carrying of cargo containers from seaports. The other firm with promising results was Magnemotion, a Massachusetts firm led by Richard Thornton, formerly of MIT and one of the developers of the EDL Magneplane system back in the 1970s. (Thornton's partner in the Magneplane project, Henry Kolm, founded Magplane Technology, Inc., which is currently building an electromagnetic non-maglev pipeline system for transporting coal in Mongolia.) At the 2005 FTA workshop, Thornton reported on the development of their M3 (Magnemotion Maglev) system, which employs the EML attractive levitation approach, but instead of using iron-core electromagnets as in the Transrapid and HSST systems, uses neodymium permanent magnets to provide most of the upward magnetic force. The perma-nent magnets are supplemented by an electromagnet and feedback control to provide vertical stability, with a levitation gap of nearly an inch (about twice the gap of the Transrapid and HSST EML systems). The same permanent magnets also provide lateral guidance forces and interact with currents in the guideway to provide propulsion. Magnemotion built a 1/7 scale vehicle and tested it on a short track. Their team also emphasized the importance of small, lightweight vehicles (about the size of a small bus) and lightweight guideways for urban maglev systems. Both General Atomics and Magnemotion use linear synchronous motors for propulsion.

The FTA "lessons learned" report of 2009 concluded that urban maglev systems were "advanced enough, but the initial infrastructure costs are intimidating," and that "lack of an actual system in place to demonstrate projected savings in maintenance and operation costs also contribute to a reluctance to embrace the technology." However, in 2008 Magnemotion received additional FTA funding to work with Old Dominion University in Norfolk, Virginia, to demonstrate operation of a small M3 vehicle on part of an existing guideway built on the campus some years earlier as part of an unsuccessful project led by American Maglev Technology (AMT). AMT's maglev vehicle, even at low speeds, vibrated and rattled severely on the track, and they ran out of money before they could fix the problem. With AMT gone and the track and vehicle remaining on campus, the students call the unfinished maglev project "Mag Left." (AMT now has a working EML maglev vehicle operating on a test track about 2,000 feet long in Powder Springs, Georgia, and is marketing their technology to potential customers undeterred by AMT's past difficulties in Florida and Virginia.) Another university campus with plans and hopes (and FTA funding) for a low-speed maglev system is California University of Pennsylvania, which plans to use the General Atomics Inductrack system for what could eventually become a 4-mile line, called the Sky Shuttle, linking different parts of their extensive campus. The campuses of Old Dominion University and California University of Pennsylvania are not very urban, but what progress is made on those campuses will determine at least the short-term future of urban (low-speed) maglev in the US. For now, if you want to experience urban maglev, go to Nagoya and take a ride on the Linimo.

Permanent-Magnet Maglev

The magnets providing levitation in the two maglev systems in the most advanced stages of development, Germany's Transrapid (EML) and Japan's Yamanashi line (EDL), are electromagnets—iron-core copper-wound electromagnets in the German EML and liquid-helium-cooled superconducting electromagnets in the Japanese EDL. However, it is interesting that the two most promising U.S. urban maglev systems funded by the FTA, Magnemotion's M3 (EML) and General Atomic's Inductrack (EDL), each employed neodymium permanent magnets. Permanent magnets have the advantage over electromagnets that they provide magnetic fields and forces without the need for expending electrical power.

In West Berlin, there was in the 1980s a short-lived operating urban maglev system based on permanent magnets called the M-Bahn (Magnet-Bahn or magnet train). The M-Bahn was 1.6 km long on an elevated track with three stations, and it used permanent magnets on the cars to provide upward attractive forces, toward the iron track, that lifted most of the weight of the cars. Of course, as we discussed in Chapter 3, such an attractive approach is vertically unstable, but instead of using sensors and feedback to control the gap as in the Transrapid and HSST, the M-Bahn cars simply had wheels that contacted the underside of the track. The cars, like the rotors in watt-hour meters and Ahmadinejad's uranium centrifuges, were thus not fully contact-free and perhaps could be called maglift rather than maglev. Whatever you may want to call it, the M-Bahn line functioned well, running very quietly and smoothly, but after the Berlin Wall fell in 1989 and West and East Berlin were no longer separated, the Berlin transportation system was modified and the M-Bahn was dismantled in 1991. In its short operating life, however, the M-Bahn had safely and comfortably carried over a million passengers.

Several scientific supply houses offer model maglev train kits of varying price and complexity, and most achieve levitation by simple repulsion between like poles of permanent magnets, which we discussed in Chapter 3 and illustrated in Figure 3. Some years ago, I visited a physics class in my local high school to talk about magnets and magnetic forces, and when I first mentioned magnetic levitation and maglev trains, the teacher made a quick trip to his supply cabinet and brought out a model maglev train he had demonstrated in an earlier class. The track contained two parallel magnetic strips mounted on a long strip of wood, and the car, about an inch wide and 6 inches long, had two parallel magnetic strips on its bottom with poles oriented so that they were repelled upward by the magnets in the track. Such a repulsive system is stable vertically but unstable laterally, so the track had plastic side-rails to prevent lateral displacements, much as we used a pencil to stabilize the levitation of the magnets in Figure 3. I suspect that similar maglev train kits have made an appearance in numerous school science fairs. The Chicago Museum of Science and Industry has a popular interactive maglev exhibit based on permanent magnets and repulsion between like poles.

To use direct repulsion between like poles of permanent magnets for full-scale maglev trains, the magnets would have to be very resistant to demagnetization by opposing magnetic fields, that is, possess a high value of a magnetic property known as *coercivity*. And to be considered as practical for many miles of track, they must also be quite inexpensive. These two conditions were first reached in the 1960s with the development of ferrite magnets, which had coercivities considerably higher than those of previous permanent magnets (alnicos and steels) and were also inexpensive. The arrival of ferrites inspired Geoffrey Polgreen, a British engineer, to promote maglev trains based on repulsion between permanent magnets in his 1966 book, *New Applications of Modern Magnets*. He worked with the

British Rail Research Center to patent what they called the Magnarail system, and built a model sufficiently large to carry one person. He estimated costs for a full-scale system with 10-ton cars, which he optimistically claimed would be less expensive than for a normal railway. He suggested wheels rolling on side rails to achieve lateral guidance, and, like other maglev dreamers before him, predicted propulsion by linear motors to speeds up to 500 km/h. In the United States, Westinghouse Electric also built a similar small model levitated by repulsion between ferrite permanent magnets.

Magnets much more resistant to demagnetization than the ferrites were developed in the 1970s (cobalt-samarium) and in the 1980s (iron-neodymium-boron). And they also delivered much higher magnetic fields and were thus capable of much higher repulsive forces. (The maximum repulsive force between like poles varies as the square of the magnetic field emanating from the poles). That opened the possibility of much larger levitation gaps than was possible with the ferrites, or the use of fewer magnets to support a given load, making this approach to magnetically levitated trains more technically feasible. There was a catch, however. The new magnets were also much more expensive than the ferrites, which would make this approach more costly. And there remained the questionable practicality of miles and miles of magnetic tracks with the tendency to attract errant steel objects that might do damage to a maglev train moving at high speed with a modest levitation gap between brittle permanent magnets. However, several groups today are using or promoting maglev transport based on the repulsion between like poles of permanent magnets on the track and permanent magnets on the car, including LevX of Washington and LaunchPoint Technologies of California, presumably with neodymium magnets.

Several groups are promoting repulsion between a permanent-magnet track and a vehicle carrying liquid-nitrogen-cooled high-temperature superconductors, which in a sense are a form of permanent magnet, as long as they are kept super-cold. As discussed in Chapter 7, bulk high-temperature superconductors like yttrium-barium-copper-oxide (YBCO), when processed to produce strong flux pinning, can behave somewhat like permanent magnets. We have seen that, paired with neodymium permanent magnets, they can provide upward repulsive forces sufficient to levitate sumo wrestlers, heavy rotors in energy-storage flywheels, and model maglev trains. The Dresden-based group, in their SupraTrans Project, has extended their work on small models to the demonstration of a larger model at the InnoTrans 2004 Fair in Germany. This system, like the smaller one, uses neodymium magnets in the track in an NSN pole arrangement and field-cooled flux-pinned YBCO in the car. Their demonstration vehicle was 1.3 m long and 170 kg when empty, and was propelled along a 7-m track with a linear synchronous motor. A university group in Brazil has constructed a similar model with permanent-magnet tracks and high-temperature superconductors on the car. They call their project Maglev-Cobra because they envision short articulated cars that can navigate curves of small radii in a snake-like fashion. A Russian group has also worked on such systems. Probably the largest maglev car to date using this approach is that built at Southwest Jiaotong University in China. A vehicle 3.5 m long, weighing 530 kg when carrying five people, has reportedly carried thousands of people along a short demonstration track. They have studied different arrangements of magnets in the track, including a Halbach array. Maglev systems based on cars with nitrogen-cooled high-temperature superconductors levitating above permanent-magnet tracks seem unlikely to be used for long-distance transport in the near future, but might find some limited applications on a smaller scale.

Transrapid Trials—and Tribulations

In my earlier book, *Driving Force* (1996), I wrote of Germany's plans to construct a 290-km Transrapid maglev line between Hamburg and Berlin: "Track construction is scheduled to start in 1996, and the target for operation is 2005." But published predictions for maglev

lines often go unmet, so I then added cautiously, "Maybe the Transrapid will meet this schedule, maybe not, but it looks today like the
best bet to become the world's first intercity maglev." I was certainly
looking forward to taking that high-speed maglev ride from Hamburg to Berlin, forecast to take less than an hour. I was not surprised
that various political and economic issues arose in Germany to de-
lay construction, including increasing cost estimates and decreasing
estimates of ridership and revenue. But the project remained officially in the government's plans until February 2000, when it was
finally canceled a few months before construction was to begin. The
cancellation was an embarrassment for many, and government officials and the Transrapid consortium realized that it was becoming
more and more difficult to promote Transrapid technologies in other
countries around the world if Germany itself was unwilling to make
the commitment. They therefore gathered to consider other options,
focusing on shorter and less expensive maglev projects. By 2003,
they decided that the most promising of the various possibilities was
a 37-km maglev link between the Munich airport and the downtown
central railroad station. And again plans progressed. But again cost
estimates escalated, in large part because it became clear that expensive tunneling would be required, and plans for the Munich maglev
airport link were finally scrapped in 2008.

It had not helped that there had been a serious accident on
Transrapid's Emsland maglev test track in September 2006. Two
workers in a maintenance car were on the track doing a routine in-
spection, but somehow that information was not known at the station where passengers boarded the maglev train for a high-speed
demonstration ride around the track. When the maintenance workers saw the train approaching, they survived by jumping off the elevated track, but the train could not stop in time, and twenty-three
people on the train were killed, including the engineer. In the colli-
sion, the train's low nose scooped up the maintenance car, which
ended on the roof. The train itself did not derail, since its lower car-
riage is wrapped around the T-shaped track. (In contrast, in the 1998

crash of Germany's high-speed ICE, the train derailed when a wheel fractured. There the toll was 101 dead and 88 injured.) Magnetic levitation was not the cause of the 2006 crash, but the high speed of the train, estimated at nearly 200 km/h, surely contributed to the casualty count. The much higher speeds that the Transrapid train is actually capable of would presumably have led to even more deaths. The cause was labeled as "human error," and later two Transrapid managers were fined for insufficient safety controls. German Chancellor Una Merkel visited the site immediately after the crash to show concern for the victims and was quoted as saying, "I don't see any connection with the technology. The technology is a very, very safe technology." But the general image of maglev trains had taken a severe hit, particularly in Germany. Fortunately for Transrapid, by that time many millions of passengers had traveled safely on their high-speed maglev trains in another part of the world—Shanghai. China (Figure 38).

Starting in the 1960s, the German government and German industry invested heavily in developing the Transrapid maglev train systems, and by the 1990s, Transrapid International was actively promoting their well-proven systems around the world. In the early 1990s, a Transrapid airport link in Orlando, Florida, appeared likely to become the first maglev line in the United States, but that project died by 1994, a precursor of later abandoned maglev projects, including the Berlin-Hamburg line and the Munich airport link. At first glance, the advantages of very high-speed trains seem to be strongest for long intercity lines, rather than for short links, but short links of course require a much smaller up-front investment, making them an easier commitment for government and industrial decision makers. And it may be easier to make such commitments in a nondemocratic country where public support is not usually an important part of the equation. Whatever the reasons, in the summer of 2000, just a few months after the Berlin-Hamburg maglev project was canceled, the city of Shanghai and Transrapid International signed an agreement for a feasibility study for a 31-km maglev line between

Figure 38. Transrapid maglev train connecting Shanghai to Pudong International Airport. Photo courtesy of Transrapid International GmbH & Co. KG, © Fritz Stoiber, photographer.

Shanghai's Pudong International Airport and a city train station. The construction contract was signed the following January.

Construction of the dual-track elevated guideway started soon thereafter, segments of the trains began to arrive from Germany, and on New Year's Eve 2002, the official maiden trip of the world's first commercial high-speed maglev line carried the Chinese prime minister, the German chancellor, and a collection of lesser VIPs on a high-speed ride. Later, a demonstration run reached a maximum speed of 501 km/h, and regularly scheduled service, with a maximum speed of 430 km/h, began in December 2003, one year after the VIP trip. The 30-km trip takes less than 8 minutes, with the average speed from departure to arrival a little more than half the maximum speed. (The train spends much of the first half of the trip accelerating, and much of the second half decelerating.) To add to the adventure, the current speed is displayed in each of the cars, but the view out the windows provides the best impression of the high speed of travel. The ride itself is said to be very smooth. By July, a million passengers had made the trip, and by the end of 2005, the total had reached 5 million.

I have not yet made the trip to Shanghai. That would be a very long flight for me to take if only to experience the brief thrill of an eight-minute high-speed train ride. But in 2005 in Slate.com, Henry Blodget presented a vivid account of the experience:

> Inside the car, you expect to see seat belts and shoulder harnesses (for all the good they would do in a derailment or collision at one-third the speed of sound), but, instead, find only normal seats. The doors shut, and the train accelerates like a skyscraper elevator, si-lently, smoothly, and rapidly, and by the time the last car leaves the station you already seem to be going 50 miles per hour. Four minutes of gravity-simulator-style acceleration later, in which the taxis on the parallel highway lose ground slowly, then quickly, then disappear as fast as if they were parked and you were whipping by at 220 miles per hour, you reach the peak speed of 270 miles per hour for the tiny 20-mile run.

Blodget's observation from the speeding Shanghai maglev that taxis driving on the parallel highway "disappear as fast as if they were parked" contrasts markedly with my observations out the train window in Connecticut of cars passing the Acela, America's "high speed train." Blodget did report some side-to-side shaking of the car at maximum speed, and "a pop and a blur as the maglev headed in the opposite direction blasts past at an aggregate speed of 534 miles per hour, approaching that of a 747 at 35,000 feet." But later in the ride, "when you've slowed to a mere 150 miles per hour, you feel as though you are strolling." When you reach the Longyang Road station after your eight-minute ride from the airport, there you find a museum describing the history and technology of maglev trains.

So the world's first commercial high-speed maglev line was not built in Germany, Japan, or the United States, but in China, which had not significantly participated in the twentieth-century research and development stages of maglev technology. Is it a success? It runs on time, and many millions have by now traveled the Shanghai maglev line comfortably and safely. The only accident was a 2006 fire as the train arrived at the terminal, but it was soon out and no one was hurt. Daily ridership, claimed to be about 20,000, is well below projected numbers, partly because of the high cost of tickets (about $10 for a round trip) and partly because the terminal is far from the city center, and most passengers then have to take the subway or a cab to their final destination. Although most take the train simply to get to and from the airport, many tourists and others take it simply for the thrill, the equivalent of exciting amusement-park rides, most of which are also only a few minutes long. (The maximum speed of the Shanghai maglev is more than twice the maximum speed of Kingda Ka, currently the world's fastest amusement-park ride. Of course, on the Kingda Ka, you are in an open car and feel the speed more directly from the impact of the air. Perhaps Transrapid should consider offering some open cars—with seat belts—to attract more thrill-seekers.) Maglev proponents claim that the revenues from ticket sales cover the maintenance and operating costs of the Shanghai airport shuttle, but ticket revenues have made little dent in the construction costs, which are estimated to have been about $1.2 billion. Kevin Coates, executive director of NAMTI (the recently formed North American Maglev Transport Institute), feels that one of the major lessons from the Shanghai maglev system is its low maintenance costs. In contrast, he argues, high-speed wheel-on-rail systems require extensive maintenance of the tracks to repair the pounding they receive and the wear and tear of friction when wheels meet rails at high speeds.

It is fair to note that the name of the Shanghai Transrapid maglev translates as "Demonstration Operation Line." As a Siemens vice president said when the contract was signed, "Transrapid views the Shanghai line, where the ride will last just eight minutes, largely as a sales tool. This serves as a demonstration for China to show that this works and can be used for longer distances, such as Shanghai to Beijing." Fair enough, but construction started in April 2008 on a 1,318-km high-speed rail line between Shanghai and Beijing, and it will be a wheel-on-rail line, not a maglev line. It is expected to have a top speed of 350 km/h and cut the train travel time between China's two major cities from 10 hours to 4 hours.

In 2009, *Fortune* magazine carried an article entitled "China's Amazing New Bullet Train (It Leaves America in the Dust!)" But the topic was not the Shanghai maglev airport link. It was China's commitment to spend, by the year 2020, $300 billion on more than 25,000 kilometers of new track for high-speed trains, including the new Beijing–Shanghai line. And they will be largely non-maglev wheel-on-rail lines. The program started in 2005, and the first result was the Beijing–Tianjin line that went into operation in August 2008, in time for the Beijing Olympics. That train covers the 114-km route in about half an hour. The world economic downturn soon thereafter encouraged China to accelerate its high-speed-rail program, as an economic stimulus and to provide needed jobs to counteract losses in its export businesses. In 2009 alone, China invested $50 billion in high-speed non-maglev trains. In December of that year, a second and longer high-speed wheel-on-rail line went into operation between the southern coastal city of Guangzhou and Wuhan, deep in the interior. That train covers over 900 km in less than three hours, despite several stops along the way.

Although most of China's new high-speed rail lines are wheel-on-rail, commercial maglev may not be forever limited to the Shanghai airport shuttle. In 2006, Transrapid signed an agreement for a 169-km maglev line running southwest from Shanghai to Hangzhou. That project was put on hold for several years, but in March 2010, the Chinese Ministry of Railways announced that this maglev line had been given the green light. In the following October, a new high-speed wheel-on-rail line between the two cities began opera-tion, but plans for the maglev line remain on the books. Perhaps by the time you read this, construction will have begun.

Sin Express to Vegas

When Barack Obama assumed the U.S. presidency in January 2009, his first action to tackle the deep recession he inherited was a giant stimulus bill, officially called the American Recovery and Reinvest-ment Act (ARRA). In February, Obama spoke before a joint session of Congress to explain the bill, and after his talk, Louisiana Gover-nor Bobby Jindal appeared on TV with the official Republican re-sponse. When he gave his assessment that the bill "is larded with wasteful spending," he surprised many with the example he gave: "$8 billion for high-speed rail projects, such as a 'magnetic levita-tion' line from Las Vegas to Disneyland." "The 311-mph train," said the *Washington Post,* "could make the trip from Sin City to Tomor-rowland in less than two hours, according to backers." "Billions of dollars for a sin express train from Los Angeles to Las Vegas. Nec-essary? I don't think so," said one congressman. Liberals pounced on Jindal and other critics, pointing out rightfully that nowhere in the stimulus bill was there any mention of a maglev line between Las Vegas and Disneyland. The $8 billion directed at high-speed rail, about 1% of the total funding in ARRA, was to go to high-speed rail projects to be chosen by the Department of Transportation from proposals submitted from around the country.

But Governor Jindal did not just invent the idea of a maglev line between Las Vegas and Disneyland or Los Angeles. The idea was of long standing. It had originated in Las Vegas in the early 1980s, when city officials and hotel owners began to worry about losing business to the new casinos in Atlantic City, and thought a high-speed train from Southern California might transport throngs of high-spending customers to their tables. Prominent Las Vegans sampled high-speed rails in Europe and Japan, and Mayor Bill Briare became especially enchanted with the potential of maglev. "Can't you imagine," he said in 1984, "a fast, sleek, beautifully designed train, traveling at very high speed, gliding across the desert from Los Angeles, carrying hundreds of happy passengers in air-conditioned comfort. Inside there is hardly any noise, no nostalgic clickety-clack, clickety-clack of steel wheels on iron rails." He gave the floating maglev train an appropriate Vegas-style compliment: "It has pizzazz."

An international conference on maglev was held in Las Vegas in 1987 at which most of the papers dealt with the German and Japanese EML and EDL systems, Berlin's M-Bahn (Las Vegas was then considering a downtown M-Bahn system), or specific technical issues. But three papers reported the results of several years of study of the potential for maglev service between Las Vegas and Southern California. Although all three were positive about the potential for high-speed-rail along this corridor, one paper noted that "operational reliability is a major factor in commercial feasibility," and felt that the French TGV wheel-on-rail system had already demonstrated by 1987 more operational reliability than any of the maglev systems, a forewarning of the competition a maglev line would later face.

Congress's Transportation Equity Act for the 21st Century, passed in 1998, kicked off both the FTA urban-maglev program, discussed earlier, and the FRA's Maglev Deployment Program, a follow-up to their 1990 National Maglev Initiative. The word "Deployment" in the new program's name indicated that the FRA's goal was to proceed beyond planning to real construction of a maglev system in the United States, conservatively aiming first at putting a "relatively short" (30–50 mile) system into service. It was to conduct a "competition to select a project for the purpose of demonstrating the use of Maglev to the general public." Backers like Senator Moynihan felt that a short demonstration line would impress the U.S. public with the reliability (and pizzazz) of maglev trains and eventually lead to intercity maglev. From eleven initial applications, the program selected seven different states for planning grants, and in 2001 chose two of the seven proposals for further funding—Pennsylvania and Maryland. Among the five rejected alternatives was one from the state of Nevada—a 35-mile corridor linking downtown Las Vegas with Primm, a tiny town on the Nevada–California border. Located mostly within the right-of-way of Interstate Route 15, the corridor "traverses an area of sparse development and gentle topography." (There is very limited demand for travel from Las Vegas to tiny Primm, but this was to be the first stage of an eventual link between Las Vegas and Anaheim, home of Disneyland.) The FRA provided most of the funds to the two selected projects, but also made some funds available to Nevada and the other losers to continue the "pre-construction planning" of their projects. An FRA chart of funding for the seven state maglev projects through fiscal year 2005 shows $20 million to Pennsylvania and $13 million to Maryland, the two winners of the early competition. Of the five losers, however, the Nevada project received the most federal money—almost $9 million. Not bad for a loser.

A September 2005 FRA report to Congress on the "Costs and Benefits of Magnetic Levitation" listed four U.S. maglev projects still under consideration: Baltimore/Washington, Pittsburgh, Las Vegas/Anaheim, and Los Angeles Airport to Riverside. However, it also noted that of the various options for developing high-speed trains—upgrading existing lines, new high-speed wheel-on-rail lines (like TGV), and maglev lines—"maglev technologies of today are the most expensive . . . in terms of up-front investment," citing estimated capital costs from $40 to $100 million per mile. Many pages of cost-benefit analyses end with a generally negative view of maglev compared to the cheaper alternatives, although it noted that in special cases, maglev could be financially competitive with new high-speed wheel-on-rail lines.

A 2006 paper by promoters of the Vegas–Anaheim maglev line noted that by 1991 it had selected Transrapid technology, that it had received continued funding from the FRA Maglev Deployment Program for many years, and that in 2005, Congress passed a new transportation act allocating $45 million "to initiate deployment of the Las Vegas to Primm project segment." That segment was to be

Phase 1 (estimated cost $1.3 billion), Phase 2 would be a short maglev line between Anaheim and Ontario ($2.6 billion), and Phase 3 a 201-mile link between Ontario and Primm, completing the full 260-mile connection between Sin City and Mickey Mouse, for an estimated total cost of a mere $12.1 billion.

Senator Harry Reid of Nevada had long championed this maglev project, noting that one-third of Las Vegas visitors come from Southern California, and he had played a strong role in steering money toward it over the years. But in 2009, he switched his support to the DesertXpress, a proposed 190-mile line using "proven, off-the-shelf European steel wheel on rail high-speed trains" between Vegas and Victorville, a city outside of Los Angeles on the southern edge of the Mojave Desert. (Could the capital X in the name be meant to connect the train's image with the X-rated aspects of Vegas, making it another "sin express"?) Maximum speed is to be 150 mph, about half the maximum speed projected for the maglev line. The estimated cost for DesertXpress is $4 billion, but backers claim that most of the money will come from private sources. Reid's switch may have been influenced by a 2008 consultant's report done for the Southern California Logistics Rail Authority. Based in large part on comparisons with the canceled Berlin–Hamburg and Munich maglev projects, the report concluded, "when compared to Maglev, the high speed rail DesertXPress project is clearly the most practical and viable alternative for the corridor." A press release in the spring of 2009 announcing the FRA's signing of the required Environmental Impact for DesertXpress claimed that the project "could break ground early next year." Magnetic levitation may have more "pizzazz" than wheel-on-rail, but sometimes pizzazz is not enough.

Maglev Deployment Program and ARRA

The Maglev Deployment Program passed in 1998 provided $60 million for preconstruction studies and potentially $950 million for actual construction. Although the funds for the planning studies were spent on several projects over the years, none of the U.S. mag-lev projects has yet proceeded to deployment. As noted earlier, the initial FRA process selected maglev projects in Pennsylvania and Maryland. The Pennsylvania project was to run from the Pittsburgh airport to downtown Pittsburgh and beyond to Monroeville and Greensburg—a total run of 54 miles at an estimated cost of $3.5 billion, with an estimate of $1.7 billion for the initial link between the airport and downtown. The Maryland project proposed linking Baltimore and Washington, D.C., with an intermediate stop at the Baltimore–Washington airport—a total run of 39 miles, with an estimated cost of $3.7 billion. Environmental Impact Statements were drafted and public hearings held. Whereas the National Maglev Initiative of the 1990s encouraged the development of U.S. technology, and the four funded groups in that program all based their systems on superconducting magnets, the Pennsylvania and Maryland projects, like the Las Vegas–Primm project and other current U.S. maglev projects, now all propose using Germany's Transrapid EML technology, which uses copper-wound, iron-core electromagnets. With Transrapid's Shanghai maglev trains carrying many thousands of paying passengers every day, opponents no longer can accurately call maglev an "unproven technology." But most of them still do.

Pennsylvania senator Arlen Specter, Republican-turned-Democrat, temporarily replaced Patrick Moynihan as the Senate's chief champion of maglev and was a prominent speaker at the international Maglev2008 conference in San Diego. (The December conference, the twentieth of the series, drew 200 participants from sixteen different countries.) In September 2009, Specter and his Senate colleague Bob Casey announced that the Pittsburgh maglev project, which already had received over $20 million in federal funding over the years, would receive an additional $28 million from the FRA "to complete pre-engineering" and other necessary preconstruction steps. Environmental Impact Statements and required public hearings have already been held, but the project still seems far from actual construction, which would require billions, not millions. Further planning funds have also been given to a proposed maglev line linking Atlanta and Chattanooga (a maglev Chattanooga Choo Choo), but that and the Las Vegas–Primm, Baltimore–Washington, and other proposed U.S. maglev projects all currently seem far from imminent deployment.

But how about ARRA, the giant stimulus package passed in February 2009, and its $8 billion for high-speed rail, the "wasteful spending" attacked by Governor Jindal and others? The bill also suggested adding $1 billion in each of the next five fiscal years for a total of $13 billion in federal funds as a "down payment" to develop high-speed rail in the United States. The short-term (several years) goal of the funding, along with other infrastructure spending, was primarily job creation, but it also coincided with the long-term U.S. goals of reduced dependence on foreign oil and decreased emission of greenhouse gases. The bill does not specify whether the funded high-speed train projects should or should not be maglev, but clearly any maglev proposals would be competing with non-maglev proposals. And to spur short-term job creation, the ARRA-funded projects were to be "shovel ready," which cannot yet be said of any of the maglev projects.

In April 2009, the FRA sent Congress a report on its "Vision for High-Speed Rail in America." It noted that "after 60 years and more than $1.8 trillion of investment, the United States has developed the world's most advanced highway and aviation systems" in contrast to decades of "relatively modest investment in passenger rail." As their figures show, federal investment in highway and air was already substantial in the 1950s and 1960s, but that even modest federal investment in rail did not begin until the 1970s. Now, to meet twenty-first century challenges, President Obama proposed "investing in an efficient, high-speed passenger rail network of 100–600 mile intercity corridors that connect communities across America." "Railroads were always the pride of America, and stitched us together," said Obama in April, "Now Japan, China, all of Europe have high-speed rail systems that put ours to shame."

By the summer of 2009, more than forty states had submitted pre-applications for high-speed-rail ARRA funds, totaling over $100 billion. When the accepted projects were finally announced in January and October 2010, the awards went to a number of high-speed rail projects across the country, with the major funds going to California, Florida, and the Chicago area, but all for wheel-on-rail. And no Sin Express to Vegas. For the immediate future, Americans who want to experience the "pizzazz" of a high-speed maglev train ride will have to travel to Shanghai.

Keeping It Up

Up with Magnets!

Magnets do lots of important things in modern technology, but in this book we have focused on one—the use of upward magnetic forces to oppose the pervasive downward forces of gravity, the magic of *magnetic levitation*—maglev. In some cases, like the Levitron and floating desk toys, the aesthetic appeal of levitation was primary. We are so accustomed to the steady pull of gravity that it is undeniably cool to see an object floating in space with no visible means of support. However, in most of the cases we discussed, the main goal of maglev was not aesthetic, but very practical—to reduce friction by allowing contact-free (or nearly contact-free) motion, thereby "fighting friction by fighting gravity." In our trip through the world of maglev, we have touched on many different areas of science and technology, and it now will be helpful to review where we've been.

One important point we made early on was that magnetic forces decrease rapidly with distance of separation. Because magnets are dipoles rather than monopoles, their fields and forces decrease with distance even faster than the inverse-square laws of gravity and electrostatics. So although magnetic forces can be very strong, even strong enough to lift multi-ton objects, they can't lift them very far. Among the many maglev examples we discussed in this book, the largest gap between the levitated object and the levitating magnet was 1.5 meters for MIT's large and powerful superconducting dipole carrying a circulating current of over a million amps, their "snowball in hell." A few other examples had levitation gaps of several inches, like the Japanese maglev trains using high-field superconducting magnets to achieve EDL (electrodynamic levitation) and Martin Simon's magnet platform inside the pig's abdomen. Most other examples of maglev had levitation gaps of less than an inch, like the Transrapid trains and most magnetic bearings, and some had far less, like the ASML photolithography stages. So magnetic levitation seems unable to explain the higher flights of Mary Poppins, Harry Potter, Peter Pan, the Flying Nun, et al. They all must have access to some other source of magic.

Another important point we discussed was Earnshaw's rule: Stable, fully contact-free levitation can't be attained with only the forces between static magnets. The levitated magnet will be unstable for displacements in at least one (often more) of the six degrees of freedom. For displacements in at least one degree of freedom, there are no restoring forces, no minimum in potential energy. Repulsion between like poles of magnets is vertically stable, but horizontally unstable (so we needed a pencil in Figure 3). Attraction between unlike poles is horizontally stable but vertically unstable.

We first found we could get around the Reverend Earnshaw with a *spinning magnet, the Levitron* toy. There levitation was produced by upward repulsion between like poles of permanent magnets, but it was the spin of the top magnet that produced the stability. Earnshaw's rule applied to static magnets, and the spinning top was not static.

Another option, of much more general use, was to use Faraday and Henry's discovery, *electromagnetic induction*. We found we could achieve stable levitation with eddy currents induced by *relative motion* between magnets and conductors. Examples included those magnets floating above the groove between two spinning rods of aluminum or copper (stationary magnets, moving conductors), and the Air Force (Figure 33) and NASA rocket sleds and EDL trains (stationary conductors, moving magnets). Earnshaw's rule doesn't apply where there are motion-induced eddy currents. We could also do it with repulsive eddy currents induced by pulsed or *alternating fields* (AC electromagnets). Examples included that levitated frying pan in Figure 13, the aluminum plate that Bachelet lifted through his goldfish bowl, and the levitation melting of reactive metals. Here the changing magnetic fields that induced eddy currents resulted from changing the currents in electromagnets rather than from relative motion.

We also learned that Earnshaw's rule doesn't apply to *diamagnetic materials,* materials repelled by magnetic fields—like water, frogs, or pyrolitic graphite. Water and frogs, like you and me, are only very weakly diamagnetic, so we needed a very high magnetic field and a high field gradient—the magnetic force depends on field gradient!—to produce the Moses effect (separating water, like Moses parted the Red Sea) in a horizontal superconducting magnet. And we needed a high field and high field gradient in a vertical superconducting magnet to produce froglev—a flying frog (Figure 17). The diamagnetism of pyrolitic graphite is much stronger than that of a frog, so here we could levitate graphite flakes even above a square array of neodymium permanent magnets, but only to heights of a millimeter or so. Diamagnetic materials were also shown capable of stabilizing levitation where otherwise it would be unstable, as noted above. A permanent magnet levitated by upward attraction to the opposite pole of another permanent magnet is vertically unstable, but if you put diamagnetic materials near the lower magnet, they provide a local repulsive force that stabilizes the levitation. We could do that with a couple of graphite slabs (Figure 15), or even, in special cases, with a pair of fingertips (Figure 16). Earnshaw would probably have been amused by that, but it didn't break his rule, which doesn't apply when diamagnetic materials are involved.

And Earnshaw's rule doesn't apply to *superconductors.* Type I (low critical field, low critical temperature) superconductors are super-diamagnetic, and we could stably levitate permanent magnets above them if the superconductor was bowl-shaped to give the magnet lateral stability. That's how Arkadiev did it way back in 1945 with lead cooled to liquid-helium temperature (4.2K), a demonstration that impressed Shoenberg (and later me, when I first saw it at GE) with the aesthetic appeal of magnetic levitation. Type II (high-field) superconductors, on the other hand, are not fully diamagnetic—they allow some field penetration. And flux pinning within the Type II superconductor allows a magnet to levitate stably above even a flat superconductor, a demonstration that became very popular with the development of high-temperature superconductors like YBCO (yttrium–barium–copper–oxide) that require cooling only to liquid-nitrogen temperature (77 K). And if the Type II superconductor has sufficient flux pinning, its internal supercurrents can produce self-stabilizing levitation above, below, or even alongside a magnet (Figure 19). Flux-pinned YBCO, coupled with neodymium magnets, has been used to levitate sumo wrestlers (Figure 18) and to provide passive bearings in energy-storage flywheels and in model maglev cars above permanent-magnet tracks, including one car in China large enough to carry several people.

So you can use exceptions to Earnshaw's rule and produce fully contact-free levitation with a spinning magnet, with eddy currents, or with a diamagnet or superconductor. Another option is just to learn to live with it and accept a device in which the forces between permanent magnets don't produce fully contact-free levitation, but still bear most of the weight, and therefore can greatly decrease friction. We did that first with the Revolution toy (Figure 4), in which the forces between the disc magnets in the rotor and the triangular magnets in the base (Figure 6) provide stability in four degrees of freedom (two translational, two rotational). The rotor is unstable only with regard to displacements along its axis, but physical contact of the rotor tip with a glass plate limits that displacement. We also used that general approach in the watt-hour meter (Figure 26), uranium centrifuge (Figure 27), some flywheels and wind turbines (Figure 29), and Berlin's M-Bahn urban maglev. The moving parts in those machines were not completely contact-free, but with much of their weight supported by magnetic forces, they were part of very useful, low-friction machines. If you prefer to reserve the term *maglev* for fully contact-free levitation, these examples were *maglift*, not maglev.

The most potent way to get around Earnshaw's rule and produce contact-free levitation is the servo approach, *using sensors and feedback* circuits to produce stabilization in the needed directions. That's what we did to stabilize the floating globe on my desk at MIT, the floating plane models in wind tunnels, and many other levitated objects in Chapter 8 and subsequent chapters. With Hall sensors measuring local magnetic field, or optical, capacitive, inductive, or other types of position or gap sensors feeding their data to feedback circuits, we achieved fully contact-free levitation in lots of situations. That's what we used to levitate that giant superconducting dipole at MIT (Figure 24), those microbots at Waterloo, and the magnet platform inside the abdominal cavity of a live pig (Figure 25). New desk toys like the Levitron AG (Figure 21), and the Crealev levitation modules that levitated those twelve red balls in Jane Philbrick's "Floating Sculpture" (Figure 22) are an improvement over earlier desk toys. All the magnets, sensors, and feedback circuits producing stable levitation in these devices are in the base below, and thus there is nothing above the floating object, which increases their aesthetic appeal.

Active magnetic bearings—bearings that use feedback, as opposed to passive bearings that don't—use servo systems for levitation in flywheels, blood pumps (Figure 28), and many other rotating machines. And we use them in nonrotating maglev devices, including flying broomsticks (Figure 30), flotors in haptic systems (Figure 31), photolithography stages to manufacture integrated circuits, and EML maglev trains in commercial service, like the high-speed Transrapid airport shuttle at Shanghai (Figure 38) and the low-speed Linimo line outside Nagoya. Systems employing active bearings generally require a fail-safe backup to protect against damage from possible power outages.

The different types of maglev systems we have discussed employ a variety of magnet types, including the three types shown in Figure 2: *permanent magnets* (usually iron–neodymium–boron), *iron-core electromagnets* (for variable and controllable high magnetic fields), and *air-core electromagnets* (high-field superconducting electromagnets, usually wound with niobium–titanium or niobium–tin, or low-field copper-wound air-core electromagnets where linear current-field behavior is important for control and precision). A fourth magnet type used in some maglev applications is *bulk type II superconductors with strong flux pinning,* usually YBCO, which act as self-stabilizing "permanent magnets" as long as they remain cooled well below their critical temperature.

For systems using feedback control, at least part of the magnetic force providing levitation is provided by an electromagnet, since its field and the resulting force can be adjusted by modifying the current in response to feedback signals. In the Transrapid and Linimo maglev trains, it is an iron-core electromagnet that provides the upward attractive force toward the track. In the Magnemotion M3 urban maglev EML system, much of the upward magnetic force is provided by neodymium permanent magnets, but they are supplemented by an electromagnet, whose modest field can be adjusted to provide stable levitation at a controlled levitation gap. Active magnetic bearings all employ electromagnets, usually iron-core.

Feedback control systems are not used in maglift systems like watt-hour meters, which use only permanent magnets, and are not needed in high-speed EDL trains stably levitated by repulsion from motion-induced eddy currents in conductors in the track. The Japanese EDL trains carry high-field superconducting electromagnets, developed in the 1960s after the discovery of Type II superconductors. The General Atomics urban maglev Inductrack EDL system instead uses neodymium permanent magnets in a Halbach array, which produces an enhanced magnetic field from one surface sufficient to produce an acceptable levitation gap at modest speeds. (All EDL trains need relative motion for levitation and so need wheels to support them until they reach sufficient speed to "take off.") Once eddy currents are induced in the conductors in the track, the conductors in the track become electromagnets. And of course the linear motors that provide propulsion in both EDL and EML trains also involve electromagnets, as do the motors that move and rotate other maglev and non-maglev systems.

In some cases, like photolithography stages and the flotors in the maglev haptics system, levitation and propulsion are produced and controlled by forces between neodymium permanent magnets and air-core (iron-free) copper-wound electromagnets. Although iron cores in electromagnets increase the magnetic field produced by a given current (Figure 2), the magnetic behavior of iron is not fully linear or reversible. Iron-free electromagnets provide a more reliable linear relation between current and field, an important consideration for control systems in which precision is important.

Magnetic Personalities

And of course many human stories are also intimately associated with all these technology stories, starting with the Reverend Samuel Earnshaw, a British preacher and teacher who apparently was comfortable in both science and religion, areas of thought often viewed as being in conflict. He published many papers in physics and mathematics as well as numerous sermons and articles on religious topics. But he remains known today primarily through one thing—his famous theorem about the instability of electromagnetic forces, published in 1842. Today the theorem most commonly appears in discussions of magnetic levitation, but it was originally applied primarily to electrostatic charges and fields in a paper entitled "On the Nature of the Molecular Forces Which Regulate the Constitution of the Luminiferous Ether." His conclusion about instability is usually called Earnshaw's theorem, but for the conditions he considered, the mathematical proof is certain and follows directly from the spatial dependence of electric and magnetic fields and forces. So rather than calling it a theorem, which to many nonmathematicians implies uncertainty and doubt, it could perhaps better be called Earnshaw's rule, Earnshaw's law, or even Earnshaw's fact.

Jumping ahead to the 1990s, a simple maglev toy, the Levitron, motivated many prominent scientists to analyze in detail the subtle physics and mathematics underlying its spin-stabilized levitation, and led to intense legal battles over patents in which several of those scientists became involved. (One of those scientists challenged the analysis of magnetic fields and forces in one of the competing patents because it was not consistent with Earnshaw's theorem/fact.) Emotions about the Levitron controversy still run high today and contribute to the intensity of new patent challenges about the recent feedback-controlled desk toy, the Levitron AG. Competing claims also play a role in the early history of two other topics we discussed, targets that have much greater impact on science and technology than the Levitron—the joint discovery of induction in the 1830s by Michael Faraday and Joseph Henry and the joint invention of integrated circuits in the 1950s by Bob Noyce and Jack Kilby. These two examples show that when important scientific or technological advances are imminent, often more than one person has the necessary insight to connect the dots.

I wish I had been around that lab in Nijmegen when Andre Geim and his colleagues first discovered that their high-field superconducting magnet could levitate water and other diamagnetic objects, including wood, plastic, cheese, pizza, and then even living creatures like grasshoppers and frogs. For those involved, it was a very exciting and enjoyable time. Several scientists had previously demonstrated diamagnetic levitation with inanimate objects and had published papers describing their results. But Geim's inspiration to demonstrate the phenomenon to the world with a flying frog led to greatly expanded interest in diamagnetism among scientists and the general public, to enhanced recognition of the Nijmegen laboratory, and even to an Ig Nobel Prize (which Geim and Berry accepted with a call for "more science with a smile"). The Japanese laboratory that demonstrated the properties of their YBCO super-conductors by using them to levitate a sumo wrestler demonstrated similar ingenuity and humor, and also drew worldwide attention. The other Japanese scientists who observed the splitting of water in a horizontal high-field superconducting electromagnet and dubbed it the Moses effect also contributed to "science with a smile."

Magnetic bearings represent the most widespread use of maglev in industry today, and there are many associated human stories, from the bold but ultimately unsuccessful attempt by Gordon Smith to paint broom handles on the fly, and the numerous patients now carrying maglev blood pumps within their chests, to the involve-ment of Ed Begley Jr. and Jay Leno in the promotion of maglev wind turbines. Of such stories, the history of uranium centrifuges, from German Gernot Zippe to Iranian Mahmoud Ahmadinejad via Pakistani A. Q. Khan, is the most fascinating, but also the most disturbing.

Pioneers Emile Bachelet and Werner Kemper each bravely and persistently pursued their dreams of maglev trains for decades. Bachelet's patents and demonstrations preceded Kemper's, but Kem-per's approach turned out to be more practical, and, unlike Bachelet, he was fortunate to live long enough to see his dreams become real-ity in the form of multi-ton demonstration maglev trains. The cre-ative ideas of Robert Goddard during his undergraduate years, ex-pressed in essays and short stories, are impressive and perhaps would have led to an early American maglev train if Goddard had not switched his imagination and energy into the field of rocketry, where he made history. Kemper's ideas led to the development of the EML attractive approach to maglev, now employed in China in the first commercial high-speed maglev line and in Japan in a commer-cial urban maglev line. The ideas of James Powell and Gordon Danby, triggered by that 1961 traffic jam at New York's Throgs Neck Bridge, led to the EDL repulsive approach to maglev, explored in Germany, the United States, and elsewhere. It was developed most extensively in Japan, where it remains Japan's choice for high-speed maglev, although Transrapid's EML attractive approach is today more popular elsewhere. (One author suggested that one source of that preference is "politicians and journalists for whom the words 'attractive' and 'repulsive' are easily interpretable as 'desirable' and 'undesirable.'") Like Bachelet and Kemper before them, Powell and Danby were motivated by their love of advanced technology, but to that motivation was added the annoyance of highway congestion. Growing airport congestion and other annoyances of airline travel also motivate many promoters of high-speed trains. Later stages of maglev train development in the United States and abroad are intimately interwoven with international competition, domestic political and economic conflicts, and intense lobbying by commercial interests, both for and against maglev. Important players in this part of the maglev story in the United States include not only engineers, industrialists, and government agencies, but also influential politicians such as the late Senator Patrick Moynihan and Senator Harry Reid (who was reelected in November 2010) and Arlen Specter (who was not). That story is still unfolding.

Ben Franklin's Vision: Levity over Gravity

Joseph Henry, co-discoverer of electromagnetic induction, is generally considered to be the most important American scientist of the nineteenth century, but the most important American scientist of the *eighteenth* century is a man best known today for other things. Benjamin Franklin was a Revolutionary politician, diplomat, businessman, author, publisher, and inventor, but also was widely recognized in his time for his contributions to the advance of science. For his important early work in electricity, the British Royal Society awarded him the Copley Medal in 1753. Electricity was undoubtedly one of the topics Franklin discussed in London with leading British scientist (and Unitarian minister) Joseph Priestley when they met there during the winter of 1765–1766. Priestley's book *The History and Present State of Electricity,* published the following year, included mention of Franklin's famous kite experiments. It also described many of Priestley's own experiments in electricity and notably suggested that the inverse-square law, applied by Newton to gravitational forces, might also apply to electric forces. That suggestion was experimentally confirmed twenty years later by Coulomb. After their meeting in the 1760s, Franklin and Priestley continued their friendship. In February 1790, a few months before his death, Franklin wrote Priestley a letter including some interesting predictions about future advances in science and morality:

> The rapid Progress true Science now makes, occasions my Regretting sometimes that I was born so soon. It is impossible to imagine the Height to which may be carried in a 1000 Years the Power of Man over Matter. *We may perhaps learn to deprive large Masses of their Gravity & give them absolute Levity, for the sake of easy Transport.* [italics added] Agriculture may diminish its Labour & double its Produce. All Diseases may by sure means be prevented or cured, not excepting even that of Old Age, and our Lives lengthened at pleasure even beyond the antediluvian Standard. O that moral Science were in as fair a Way of Improvement, that Men would cease to be Wolves to one another, and the human Beings would at length learn what they now improperly call Humanity.

Science and engineering were still progressing rather slowly in the eighteenth century, so Franklin estimated it might take a thousand years to achieve such wonders. But progress in the "Power of Man over Matter" accelerated mightily in the nineteenth and twentieth centuries, and many of Franklin's predictions have already been substantially achieved. He predicted, "Agriculture may diminish its Labour & double its Produce." and here he was overly conservative, since the productivity of agriculture has already much more than doubled. He predicted, "All diseases may by sure means be prevented or cured, excepting even that of Old Age." That's not yet fully achieved, but we have come a long way. In the 200 years since Franklin's death, medicine has learned to prevent or cure most of the diseases commonly afflicting people in his time, and life expectancy has more than doubled. About moral progress, he wished, "O that moral Science were in as fair a Way of Improvement, that Men would cease to be Wolves to one another, and the human Beings would at length learn what they now improperly call Humanity." Franklin had seen man's inhumanity to man in peace and war, both in America and Europe, and clearly doubted that our moral progress would match our scientific progress. Unfortunately, that has been the case to date, but perhaps we humans will get there in a thousand years—if we don't obliterate ourselves first.

Franklin's prediction most relevant to this book is the sentence I italicized above and here: *"We may perhaps learn to deprive large Masses of their Gravity & give them absolute Levity, for the sake of easy Transport."* This sentence may have been inspired by the first balloon flights of the Montgolfier brothers in Paris seven years earlier, and doesn't suggest that Franklin foresaw magnetic levitation, although the quote often appears in websites and books about maglev. Franklin was born a little too soon to recognize the close connection between magnetism and his own special interest, electricity. (His major connection to magnetism was as member of a 1784 commission in Paris debunking Franz Anton Mesmer's claims of healing patients with "Animal Magnetism.") When Franklin wrote that letter to Priestley in February 1790, Oersted was only 12 years old, and thirty more years would pass before his experiments showed that an electric current could rotate a compass needle. Faraday and Henry were not even born until later in the 1790s, and forty years would pass before their own discoveries of electromagnetic induction. But Franklin clearly recognized that the earth's gravity and the resulting weight of people and goods limited their "easy Transport," and that some "Power of Man over Matter" in the form of levitation would help. Two hundred years later, magnetic levitation has become one of the main methods we use "to deprive large Masses of their

Gravity & give them Levity," and we have in recent years found many uses of maglev.

As I write this in 2009, maglev has already made a significant impact on several areas of technology, particularly in active magnetic bearings, which first came on the industrial scene about thirty years ago. The modest levitation gaps in magnetic bearings are hidden inside the devices, so this form of maglev lacks the aesthetic appeal of the floating globe on my desk and similar popular gadgets. Maglev bearings are usually unseen, but by reducing friction they have improved the performance of an increasing number of rotating industrial devices, both common ones like motors, generators, turbines, and turbo-molecular pumps, and more exotic ones like flywheels for energy storage and blood pumps for assisting damaged hearts. And they already play an important role in many nonrotating devices, such as multimillion-dollar lithography stages for the production of integrated circuits. It seems likely that the use of various types of active and passive magnetic bearings will increase in coming years. One interesting type of passive bearing recently being explored is based on the forces between permanent magnets and high-temperature superconductors, usually in the form of nitrogen-cooled bulk YBCO (yttrium–barium–copper–oxide) with high flux pinning. Because the superconductors are self-stabilizing, no feedback control circuits are needed. For this reason, and because superconductivity continues to hold fascination for many, work on such bearings is likely to continue, and some limited applications will probably be found. Applications will of course not be all that limited if we ever find materials that are superconducting at or near room temperature.

Of the other maglev examples discussed in this book, maglev haptics technology, though relatively new, appears quite promising. Maglev wind turbines are also promising because they have the advantage of today's strong interest in green technology and of promotion by Ed Begley Jr. and Jay Leno. However, as with most other maglev devices, the extent of their application will depend on competition with non-maglev systems that are more familiar and less expensive. One of the most fascinating examples discussed in the book was maglev surgery in the abdomen of a living pig. However, this may remain a technology "ahead of its time" until other medical groups become committed to its study, which to my knowledge has not yet happened.

The "flying frog" demonstrations of Geim and colleagues in the late 1990s increased interest in diamagnetic levitation. In September 2009, scientists at California's Jet Propulsion Laboratory announced in *Advances in Space Research* that they were using a 17-tesla superconducting electromagnet to levitate small young mice for extended times to study the effect of reduced gravity on bone loss and other physiological changes. These experiments can complement animal studies on the space station, and by simply moving the mouse's cage from the levitation region to other positions in the room-temperature bore of their electromagnet, they can change the net downward force from zero G to as high as 2G. (When the cage is below the electromagnet, the diamagnetic force and the normal gravitational force add rather than subtract). They therefore call their apparatus a "variable gravity simulator." The mice are biologically closer to us than Geim's frog, making research results more transferable to human physiology. Geim had reported that the levitated frogs showed no signs of distress, but comic Dave Barry questioned a frog's ability to communicate distress. Mice are more capable in that department. The researchers report that the mice moved in a very agitated fashion when first experiencing levitation, but that they became acclimated quickly, and were soon eating and drinking normally. No hysterical squeaking was reported.

Ben Franklin's prediction of learning "to deprive large Masses of their Gravity & give them absolute Levity, for the sake of easy Transport" seems most relevant to the maglev application best known to the general public—the *maglev train*. This use of maglev has taken both several steps forward and several steps backward since the 1996 publication of *Driving Force*. We now have an operating high-speed maglev line in Shanghai and an operating low-speed urban maglev line outside of Nagoya, and each has carried millions of passengers. But the planned Berlin–Hamburg maglev line that seemed possible in 1996 (the German Bundestag had already approved many billions for construction) was canceled in 2000. The subsequently planned fallback project, the maglev Munich airport shuttle, was canceled in 2008, like the Orlando airport shuttle much earlier. China has made a major financial commitment to a national high-speed rail system, mostly wheel-on-rail, but also plans a maglev line between Shanghai and Hangzhou tentatively scheduled to be completed in 2014. In the United States, recent government reports on the FTA urban maglev program and the FRA maglev deployment program are rather negative on maglev, focusing on its high up-front construction costs. The Obama stimulus package included $8 billion for high-speed rail, but the money was awarded in 2010 to non-maglev systems.

High-speed intercity maglev trains levitated with superconducting electromagnets are still on the books in Japan. The Central Japan Railway Company has recently committed to building, with its own money, a 290-km high-speed superconducting maglev line from Tokyo to Nagoya at an estimated construction cost of over $50 billion. A majority of the line will be in tunnels through the mountainous region of central Tokaido. The target date for completion of this project, now termed the Tokaido Shinkansen Bypass, is 2025, to be eventually followed by extension to Osaka.

Over the years, short-run U.S. maglev projects have received FRA and FTA research funds in the millions, but a billion-dollar commitment by the United States to actually construct even a short maglev line like the Shanghai shuttle seems unlikely at the moment. And it may take many years before we eventually decide to build an intercity maglev line whisking us along at speeds over 300 mph, as envisioned by Emile Bachelet back in 1912. But the overall use of maglev in other applications is likely to continue to grow, both with increasing use of old applications, like magnetic bearings in rotating machinery, and the appearance of new applications, like blood pumps, haptics technology, photolithography, and superconducting bearings. The aesthetic appeal, pizzazz, and practicality of magnetic levitation in "fighting friction by fighting gravity" will retain its magic for many years to come.

Sources and Suggested Readings

Acknowledgments

Index

Sources and Suggested Readings

General

Chapter 10 of James D. Livingston, *Driving Force: The Natural Magic of Magnets* (Cambridge, MA: Harvard University Press, 1996) presents a very condensed view of magnetic levitation as I viewed it in 1994. The remainder of the book offers more discussion of the various "facts about the force" and a math-free overview of the long history of magnets and their applications.

The Sources section in *Driving Force* listed many books and journal articles as "suggested readings" for those who wanted to learn more about various topics covered. I will do the same here, but the Internet has progressed a lot in the years since *Driving Force,* and for many, the most convenient place to start nowadays will be Wikipedia and other online sources you can find via Google or other search engines. In addition to in-print sources, in the following I give pertinent websites for some topics, especially for cases where I feel they are not easily found via Google or Wikipedia.

Chapter 1

Jim Ottaviani and Janine Johnston, *Levitation: Physics and Psychology in the Service of Deception* (Ann Arbor, MI: G. T. Labs, 2007) is a graphic novel presenting an illustrated history and description of the Maskelyne/ Kellar levitation trick. The trick is also described and illustrated in Paul Curry, *Magician's Magic* (New York: Dover, 2003). Jim Steinmeyer, *Hiding the Elephant: How Magicians Invented the Impossible and Learned to Disappear* (New York: Carroll & Graf, 2003) is a more general treatment of sensational stage magic tricks. On YouTube, you can see "David Copperfield Flying" as well as explanations of how he does it. YouTube also offers videos of many other levitation magic tricks, both professional and amateur.

A free download of the memoirs of Robert-Houdin in English is available on the Web.

Roger Highland, *The Science of Harry Potter: How Magic Really Works* (New York: Penguin Books, 2003) uses the intense interest in Harry Potter to briefly introduce many aspects of the "magic" of modern science, including magnetic levitation.

J. M. Cohen, translator, *The Life of Saint Teresa of Avila by Herself* (London: Penguin Books, 1987), the autobiography of Saint Teresa describes, among many other things, her experiences with mystical levitation. There are numerous biographies of Saint Teresa, and one of the best is Cathleen Medwick, *Teresa of Avila: The Progress of a Soul* (New York: Knopf, 1999). Wikipedia offers an entry on "Saints and levitation" that includes a lengthy list of saints reported to levitate.

Wikipedia's page on levitation offers links to sources on acoustic, optical, electrostatic, and magnetic levitation. NASA's electrostatic levitator is described at esl.msfc.nasa.gov. For discussion of a variety of physical levitation methods, see E. H. Brandt, "Levitation in Physics," *Science,* 243, 349–355 (1989).

Chapter 2

For NASA's "reduced gravity research program" and the "vomit comet," see jsc-aircraft-ops.jsc.nasa.gove/Reduced_Gravity/index.html.

The story of the Feynman lecture on the weakness of gravity that was interrupted by a falling loudspeaker appears in James Gleick, *Genius: The Life and Science of Richard Feynman* (New York: Vintage Books, 1993).

In addition to *Driving Force,* numerous other books give an overview of magnets and magnetism. One of my favorites is L. Gonick and A. Huffman, *The Cartoon Guide to Physics* (New York: HarperCollins, 1990), which also explains other aspects of physics in a lively and entertaining way. Fred Jeffers, *Mondo Magnets: 40 Attractive (and Repulsive) Devices & Demonstrations* (Chicago: Chicago Review Press, 2007) illustrates and explains many fascinating demonstrations with magnets, including several examples of magnetic levitation. Gerrit Verschuur, *Hidden Attraction: The Mystery and History of Magnetism* (New York: Oxford, 1993) reviews the colorful history of magnetism and the researchers who contributed to it. For an extensive introduction to the science of magnetism that uses few mathematical equations, see E. W. Lee, *Magnetism: An Introductory Survey* (New

York: Dover, 1970; originally published in 1963 by Penguin Books) is very readable, though outdated in some areas. For those more comfortable with mathematics, many undergraduate textbooks provide excellent introductions to magnetism, including B. D. Cullity and C. D. Graham, *Introduction to Magnetic Materials, Second Edition* (New York: Wiley IEEE Press, 2008). Cullity's original text was long among the most popular, and Graham has provided an authoritative update. Robert C. O'Handley, *Modern Magnetic Materials: Principles and Applications* (New York: Wiley-Interscience, 1999) provides a more advanced treatment.

A good online source for an introduction to magnetism in general is Rick Hoadley, aka coolmagnetman. His site also shows many interesting experiments with magnets. Among the many other online sources worth a look is the Magnetism Tutorial of the NDT Resource Center. (Both these sources, and many others, use as symbols for magnets the traditional horseshoe shape common for steel magnets of yesteryear. But as explained in *Driving Force,* modern magnets are no longer horseshoe-shaped.) You can access college-level introductions to magnetism (and many other topics) via MIT's "open courseware" site, ocw.mit.edu, and various courses in physics or in materials science, including my own *very* condensed "Magnetic Materials Overview" via 3A08, my freshman seminar on magnets (under materials science). More extensive and more advanced information on magnetism is available through materials-science course 3.15 and physics course 8.02.

Chapter 3

The Revolution toy is marketed by Carlisle Company, Carson City, Nevada, and Carlisle refers to U.S. Patent 5182533, granted to Gary Ritts of Los Angeles in 1993, as its origin. Some Revolutions contain battery-driven LEDs that flash as the rotor spins. Several rotating "floating pens" are levitated in the same fashion.

Isaac Asimov's comments on Swift's Floating Island of Laputa appear in Jonathan Swift, *The Annotated Gulliver's Travels,* Isaac Asimov, ed. (New York: Clarkson N. Potter, 1980).

Wikipedia and numerous other online sites cover magnetic levitation in general. One good one is www.coolmagnetman.com/maglev.htm. YouTube has numerous videos of maglev demonstrations, from small laboratory demonstrations to rides on various maglev trains. A series put together by John Roman Iwaszko in the 1990s entitled "Antigravity—The Reality" offers a series of videos of a wide variety of both magnetic and nonmagnetic forms of levitation. Videos of several types of magnetic levitation can be seen via the UCLA physics web, going to lecture demonstrations and Mag Lev.

For more advanced in-print introductions to magnetic levitation in general, see John R. Hull, "Magnetic Levitation," in *Encyclopedia of Electrical and Electronics Engineering,* Vol. 11, pp. 740–747, ed. J. G. Webster (New York: John Wiley & Sons, 1999), Thomas D. Rossing and John R. Hull, "Magnetic Levitation," *The Physics Teacher,* 29, 552–561 (December 1991), and B. V. Jayawant, "Electromagnetic Suspension and Levitation Techniques," *Proc. Roy. Soc. London* A 416, 245–320 (1988). More extensive overviews in book form include B. V. Jayawant, *Electromagnetic Levitation and Suspension Techniques* (London: Edward Arnold, 1981), P. K. Sinha, *Electromagnetic Suspension: Dynamics and Control* (London: Peter Peregrinus Ltd., 1987), and Richard H. Frazier, Philip J. Gillinson Jr., and George A. Overbeck, *Magnetic and Electric Suspensions* (Cambridge, MA: MIT Press, 1974).

Chapter 4

Spin-stabilized levitation was patented by Roy Harrigan in 1983 (U.S. Patent 4,382,245), followed in 1995 by E. E. and W. G. Hones (U.S. Patent 5,404,062).

A number of scientific papers have been written on the Levitron, including: M. V. Berry, "The Levitron, an Adiabatic Trap for Spins," *Proc. Roy. Soc.,* A452, 1207–1220 (1996), M. V. Berry and A. K. Geim, "Of Flying Frogs and Levitrons," *Eur. J. Phys.,* 18, 307 (1997), M. D. Simon, I. O. Heflinger, and S. L. Ridgway, "Spin Stabilized Magnetic Levitation," *Am. J. Phys.,* 65, 286–292 (1997), and T. B. Jones, M. Washizu, and R. F. Gans, "Simple Theory for the Levitron," *J. Appl. Phys.,* 82, 883–888 (1997). The "Hidden History of the Levitron" written by Mike and Karen Sherlock, available at amasci.com/maglev/lev/expose1.html, gives their anti-Hones view of the Levitron controversy and includes numerous links relevant to the story. The Levitron home page of Bill Hones and Fascinations, Inc. is levitron.com, where modern Levitron products are shown. In the 1960s, well before the term *levitron* was associated with the spin-stabilized top, the term was applied to fusion devices that contained a levitated current-carrying ring.

Chapter 5

There are numerous biographies of Michael Faraday and Joseph Henry, including L. Pearce Williams, *Michael Faraday* (New York: Basic Books, 1965) and Albert E. Moyer, *Joseph Henry: The Rise of an American Scientist* (Washington, DC: Smithsonian, 1997). S. P. Bordeau, *Volts to Hertz: The Rise of Electricity* (Minneapolis: Burgess Publishing, 1982) covers Faraday and Henry and others who contributed to advances in electricity and magnetism. The brief poem by Herbert Mayo appears in John Hall Gladstone, *Michael Faraday* (London: MacMillan & Co., 1872) and in numerous other sources.

A video of the Alcon Eddy Current Levitator U.S. Patent 5,319,336 can be seen via the UCLA physics department's lecture demonstration page on Mag Lev.

The "jumping ring" experiment is described in numerous articles and texts on electricity and magnetism, including Chapter 4 of B. V. Jayawant's book on magnetic levitation cited earlier. In that chapter, Jayawant also describes numerous other examples of levitation by eddy-current repulsion. Levitation of an AC coil above a conducting plane is discussed by Marc T. Thompson in "Eddy Current Magnetic Levitation—Models and Experiments," *IEEE Potentials*, 19. no. 1, 40–44 (February/March 2000) and "Electrodynamic Magnetic Suspension—Models, Scaling Laws and Experimental Results," *IEEE Trans. Educ.*, 43, no. 3, 336–342 (August 2000). Levitation melting is described on ameritherm.com, and Ameritherm videos of levitation melting of copper and steel are also available on YouTube.

Chapter 6

Diamagnetic levitation pre-froglev was reported by W. Braunbek, *Z. Phys.*, 112, 753–769 (1939) and by E. Beaugnon and R, Tournier, "Levitation of Organic Materials," *Nature,* 349, 470 (1991) and "Levitation of Water and Organic Substances in High Static Magnetic Fields," *J. Phys. III France*, 1, 1423 (1991).

The Moses effect was reported in S. Ueno and M. Iwasaka, "Parting of Water by Magnetic Fields—The Moses Effect" *IEEE Trans.*, 30, no. 6, 4698–4700 (1994).

For more about flying frogs and later demonstrations of diamagnetic levitation and diamagnetic stabilization, see M. V. Berry and A. K. Geim,

"Of Flying Frogs and Levitrons," *Eur. J. Phys.,* 18, 307 (1997), Andre Geim, "Everyone's Magnetism," *Physics Today, 36* (September 1998), A. K. Geim, M. D. Simon, M. I. Boamfa, and L. O. Heflinger, "Magnetic Levitation at Your Fingertips," *Nature,* 400, 323 (July 22, 1999), M. D. Simon and A. K. Geim, "Diamagnetic Levitation: Flying Frogs and Floating Magnets," *J. Appl. Phys..* 87, no. 9, 6200 (2000), and M. D. Simon, L. O. Heflinger, and A. K. Geim, "Diamagnetically Stabilized Magnet Levitation," *Am. J. Phys.,* 69, no. 6, 702 (2001). Berry and Geim's press release on "The Physics of Flying Frogs" can be found via Berry's home page at the University of Bristol. Two good websites that discuss diamagnetic levitation and other forms of levitation are www.hfml.ru.nl/levitate.html (Nijmegen) and www.physics .ucla.edu/~msimon/ (Martin Simon at UCLA). The latter site contains several examples of diamagnetic levitation of pyrolitic graphite mentioned in the text.

Chapter 7

One of the most extensive online introductions to superconductivity is that of superconductors.org, a site that includes many links to other sites on superconductivity. David Shoenberg, *Superconductivity* (Cambridge: Cambridge University Press, 1952) was one of the first complete books on the topic and includes a picture of a magnet levitating above a dish-shaped Type I superconductor. The advent of Type II superconductors in the 1960s led to many more books on superconductivity, including my own, written with H. W. Schadler, *The Effect of Metallurgical Variables on Superconducting Properties* (Oxford: Pergamon Press, 1965). A second and greater explosion of interest and activity in superconductors followed the discovery in the 1980s of high-temperature superconductors, which are included in more recent books such as Werner Buckel and Reinhold Kleiner, *Superconductivity: Fundamentals and Applications* (Hoboken, NJ: Wiley-VCH, 2004). Magnetic levitation with superconductors is discussed there and is the focus of Francis C. Moon, *Superconducting Levitation: Applications to Bearings and Magnetic Transportation,* (New York: John Wiley & Sons, 1994) and John R. Hull, "Levitation Applications of High Temperature Superconductors," in A. V. Narlikar, ed., *High Temperature Superconductivity 2* (Berlin: Springer, 2004).

Chapter 8

Clicking on "products pages" at the Levitron website leads you to several varieties of floating globes as well as Levitrons. Numerous varieties of levitating globes are also offered at 1worldglobes.com and other sites.

Wilbur Wright's description of their wind tunnel, and a picture of a replica, can be seen at the website of the Wright Flyer Project.

For a discussion of the promise of "Magnetic Levitation in Surgery," see www.hadasit.co.il/category/magnetic-levitation-in-surgery.

Chapter 9

For a description of Ahmadinejad's public visit to the Natanz uranium enriching facility, see William J Broad's article, "A Tantalizing Look at Iran's Nuclear Program" in the *New York Times,* April 29, 2008. The fascinating history of the Zippe centrifuge is told by Broad in the *Times,* March 23, 2004, and by Jeremy Bernstein there on March 10, 2007. See also Jeremy Bernstein, *Nuclear Weapons: What You Need to Know* (Cambridge: Cambridge University Press, 2007). For more technical detail on the centrifuge, see Houston Wood, Alexander Glaser, and R. Scott Kemp, "The Gas Centrifuge and Nuclear-Weapon Proliferation," *Physics Today,* 61, no. 9, 40–45 (2008) and Stanley Whitley, "Review of the Gas Centrifuge until 1962. Part II: Principles of High-speed Rotation," *Rev. Mod. Phys.,* 56, 67–97 (1984). J. W. Beams, "Ultrahigh-Speed Rotation," *Scientific American,* April 1961, describes his pioneering work with magnetic bearings for centrifuges.

Chapter 10

David Trumper's MIT Web page on "Noncontact Processing of Fibers, Beams, Web, and Plates" offers links to technology and papers related to Conrad Smith's broom handle painting line.

Ralph Hollis's maglev haptics program has led to the formation of a company called Butterfly Haptics. Its Web page offers information on their commercial Magnetic Levitation Haptic Interface. Other information on maglev haptics can be seen via Ralph Hollis's Web page and the page of Peter Berkelman of the University of Hawaii, a colleague of Hollis. Aldous Huxley's *Brave New World* (New York: Harper and Brothers, 1939) later appeared as a Bantam Classic. The "Going to the Feelies" question appears in Chapter 3, the visit to the Feelies in Chapter 11.

In the June 23, 1996 issue of *Design News,* Brian J. Hogan described an early version of a maglev stage for photolithography. For a current maglev system by ASML, see the November 2008 article in the *EE Times* by Kenton Williston and the description of their TWINSCAN NXT:1950i system on the ASML website. The latter notes that "the innovative magnetic levitation technology allows significant acceleration and precision gain enabling the NXT:1950i to achieve unprecedented productivity and overlay performance."

The March 2008 press release by Holloman Air Force Base on their maglev test track flight, online in Air Force Print News Today, gives more technical details. Since NASA's Maglifter program has not been active for many years, most online items about the program date from about 2000. However, the current Web page of Powell and Danby's company Maglev 2000, under "maglev applications," has a brief description with helpful diagrams.

Chapter 11

Janet Bower's article "Emile Bachelet—Inventor from Mount Vernon, New York," written for the Westchester County Historical Society and available online, is the most complete description of the life of this pioneer of maglev trains. Her article is based on reminiscences of Bachelet's eldest son. For Goddard's early articles on maglev trains, see the website of Worcester Polytechnic Institute (wpi.edu).

The website of the International Maglev Board (magnetbahnforum. de) has a brief biography of Hermann Kemper and an image of his 1937 patent. Their site also has much other information about maglev trains, including the history of speed records for maglev trains in Germany and Japan, brief descriptions of maglev projects in various countries, a discussion forum, news items, and editorials. The history of Transrapid development is outlined in detail on several websites, including those of Transrapid (www.transrapid.de), Siemens (w1.siemens.com), and the Innovative Transportation Technologies site of the University of Washington, which offers a very wide view of innovative transportation systems.

The Powell and Danby U.S. Patent 3,470,828 for an "electromagnetic inductive suspension and stabilization system for a ground vehicle" was granted in October 1969. Their Maglev 2000 website, under "M-2000 team," offers biographies of James Powell and Gordon Danby. For information on Japan's high-speed EDS/EDL trains, inspired by the ideas of Powell and Danby, see the Wikipedia entry under JR-Maglev. Kolm and Thornton's article on "Electromagnetic Flight" appeared in the October 1973 issue of *Scientific American,* 17–25.

See the August 1992 *Scientific American,* pages 103–113, for a discussion of maglev history, the National Maglev Initiative, and their prediction that the Orlando airport shuttle would be the world's first commercial maglev line. The full text of the final report on the National Maglev Initiative is available online via the National Transportation Library.

In *Supertrains: Solutions to America's Transportation Gridlock* (New York: St. Martin's Press, 1993), Joseph Vranich argues persuasively for high-speed trains, both maglev and non-maglev.

In a December 2000 article in *Barron's,* Thomas G. Donlan's article, "A Marvelous New Toy" reported his experiences on the first official trip of the Acela Express. He gave it a mixed review, noting, as I did, that sometimes cars and trucks on the Connecticut Turnpike outpaced the Acela.

Chapter 12

Proceedings of the FTA's September 2005 Urban Maglev Workshop and the March 2009 report, FTA Low-Speed Urban Maglev Research Program: Lessons Learned, are available online. For details of the urban maglev programs of Magnemotion and General Atomics, see the company websites. In the December 2007 *Popular Mechanics,* John Quain discusses the General Atomics Inductrack program and describes a ride on their maglev test track. The websites of Old Dominion University and California University of Pennsylvania have information on the maglev projects on their campuses.

Geoffrey Polgreen, *New Applications of Modern Magnets* (London: MacDonald, 1966) outlines his work on trains levitated by repulsion between ferrite permanent magnets in the car and in the track.

Henry Blodget's article, "Mine's Faster Than Yours: Riding Shanghai's Maglev, the World's Fastest Train," was posted on slate.com on March 21, 2005. Bill Powell's "China's Amazing New Bullet Train (It Leaves America in the Dust)" appeared in the August 17, 2009 issue of *Fortune.* A 60-minute film on Transrapid and the construction of the Shanghai maglev line can be accessed online at www.parallaxfilm.com/promo/maglev.

Of numerous online articles on the choice between DesertXPress and maglev for a high-speed train between Las Vegas and Los Angeles, one of the most detailed is "Maglev or DesertXPress? One Could Be Your New Ride" by Lisa Mascaro for the *Las Vegas Sun,* June 14, 2009. DesertXpress has an extensive website. The competing maglev group, now called the California-Nevada Interstate Maglev Project (CNIMP) includes Transrapid, General Atomics, and American Magline Group. The report to the Southern California Logistics Rail Authority, "Maglev or High Speed Rail in the Las Vegas to Southern California Corridor" was prepared by Heiner Bente and Jens Gertsen of BSL Management Consultants from Germany and is available online. The "pizzazz" quote by Las Vegas mayor Bill Briare appears in John Tierney, "The Little Engine That Might Not," *Science,* July/August 1984, 74–84.

A CD entitled *2008 High-Speed and Maglev Trains, Magnetic Levitation Technology, High-Speed Ground Transportation (HSGT), FRA Development Program,* produced by Progressive Management and available via Amazon and other sources, includes numerous FRA reports, including several on the Maglev Deployment Program and links to the pages of various U.S. maglev projects.

As background for the high-speed rail portion of ARRA, the 26-page FRA report "Vision for High-Speed Rail in America" was presented to Congress in April 2009. A two-page "Highlights of Strategic Plan" is also available online.

Chapter 13

Ben Franklin's 1790 letter to Joseph Priestley appears in numerous places online, including in Nathan Gerson Goodman, *The Ingenious Dr. Franklin: Selected Scientific Letters of Benjamin Franklin* (Philadelphia: University of Pennsylvania Press, 1974) and the digital edition of *The Papers of Benjamin Franklin* by the Packard Humanities Institute. Steven Johnson, *The Invention of Air: A Story of Science, Faith, Revolution, and the Birth of America* (New York: Riverhead, 2009) is an interesting brief biography of Joseph Priestley.

Acknowledgments

For this book, I have relied heavily on experts in various fields of magnetic levitation. One who was particularly helpful is Martin Simon of UCLA, an expert in a wide variety of maglev topics, including the physics and history of the Levitron, diamagnetic levitation and stabilization, and maglev surgery. Also very helpful was David Trumper of MIT, who has worked in maglev applications for many years and who patiently helped clarify many aspects of the field to me. Both Martin and David suggested topics for inclusion and provided several illustrations and helpful critiques of some of my chapters.

Darren Garnier of Columbia, a visiting researcher at MIT, kindly hosted me on a lengthy visit to his remarkable Levitated Dipole Experiment, tutored me on many technical details, provided an illustration, and critiqued my brief description of LDX. Jane Philbrick, a visiting artist at MIT's Center for Advanced Visual Studies, and Katherine Chu, a materials-science student working with her, met with me on several occasions to discuss Jane's plans that evolved into her maglev "Floating Sculpture, 2008–09," which she exhibited at the Skissernas Museum in Lund, Sweden. I am grateful to her for keeping me in touch with her progress and problems, for letting me examine her Crealev levitation module, and for numerous interesting conversations about the intersections between art and science.

As I sought to learn more about the many and diverse technical areas impacted by magnetic levitation, I used e-mail, and sometimes phone, to contact other experts in each field. I was very pleasantly impressed that nearly all of them were very willing to help me, a total stranger to them, by answering my many questions. Among those who helped me learn about maglev systems in wind tunnels were Eugene Covert, a retired MIT professor who was a pioneer in the area, Colin Britcher of Old Dominion University and Princeton, Alexander Smits of Princeton, Hiroshi Higuchi of Syracuse University, Chin Lin of National Cheng Kung University, Taiwan, Pete Jacobs and David Dress of NASA Langley, and Mike Worthey of the Wind Tunnel Connection. Bill Hones of Fascinations, Mike and Karen Sherlock of Levitation Arts, Rob Jansen of Crealev, and Manaz Ganji of Simerlab helped me with the topics of the Levitron and maglev desk toys.

It had been many years since my own research in high-field superconductors, and David Larbalestier of Florida State University and Alex Malozemoff helped bring me up to date on recent developments there. For magnetic bearings and their various applications, I am indebted for the help of Paul Allaire of the University of Virginia, Keith Field of Pentadyne, Chet Lyons of Beacon Power, Scott Kemp of Princeton, John Hull of Boeing, Charles Farabaugh of SKF, Bruno Wagner of S2M, Reto Schöb of Levitronix, James Soeder of NASA Glenn, and Jim Dill of Foster-Miller. Ryan Young was very conscientious in hunting down answers to my numerous questions about ASML's maglev photolithography stages. Others who were very cooperative and helpful in other maglev areas included Behrad Khamesee of the University of Waterloo, Donald Ketchen of General Atomics, Jim Rowan of Enviro-Energies, Yuanming Liu of the Jet Propulsion Lab, and Ralph Hollis of Carnegie Mellon, pioneer in maglev haptics.

Especially helpful with comments and suggestions on my chapters on maglev trains was John Harding, chief scientist for maglev development at the Federal Railroad Administration until his retirement and a professional for the International Maglev Board. Dick Thornton kindly hosted me on a visit to his Magnemotion plant, educated me on the promise of urban maglev, and provided me with copies of several of his recent published papers. Others who helped me learn more about developments in maglev trains were Kevin Coates, co-founder of the International Maglev Board, Alan Rao of the Volpe Center of the Department of Transportation, Jim Fiske of LaunchPoint, Larry Blow of Maglev Transport, and Bruce Armstrong of Magplane Technology.

And as always I am grateful for the continued love and support of my wife Sherry Penney and my daughters Joan, Susan, and Barbara. Barbara, a professional photographer, also assisted by traveling to Albany to photograph a church window for Figure 11.

Figure 1. Courtesy Library of Congress Prints and Photographs Division, via WikiMedia Commons

Figure 5. Figure is by LeFebvre in a French edition of *Gulliver's Travels* published in 1797 and reproduced in Isaac Asimov, *The Annotated Gulliver's Travels* (New York: Clarkson N. Potter, 1980).

Figure 8. Courtesy of Fascinations, Inc.

Figure 9. Courtesy of Nate Harler

Figure 11. Photo by Barbara Livingston

Figure 13. Photo by Donna Coveney; reprinted with permission from the MIT news office

Figure 14. Courtesy of Martin Simon of UCLA

Figure 15. Courtesy of Martin Simon

Figure 16. Figure obtained from Martin Simon, reprinted with permission of Andre Geim, owner of the fingers

Figure 17. Reproduced with permission of Andre Geim, University of Manchester

Figure 18. Reproduced by permission of Nihon Sumo Kyokai (Japan Sumo Association) and ISTEC

Figure 19. Reproduced with permission of Yoshihiko Saito of the Osaka Science Museum

Figure 20. Courtesy of Fascinations, Inc.

Figure 21. Courtesy of Fascinations, Inc.

Figure 22. Courtesy of artist Jane Philbrick and Skissernas Museum, Lund, Sweden. Photo by Christina Knutsson.

Figure 23. Courtesy of NASA Images

Figure 24. Courtesy of Darren Garnier of Columbia, visiting scientist at MIT

Figure 25. Photo by Martin Simon, published with permission of Yoav Mintz

Figure 26. Adapted from R. J. Parker, *Advances in Permanent Magnetism* (New York: Wiley, 1990); reprinted by permission of John Wiley & Sons, Inc.

Figure 27. Courtesy of Iranian president Mahmoud Ahmadinejad, who made the figures public via his website, www.president.ir.

Figure 28. Courtesy of Terumo Heart, Inc.

Figure 29. Courtesy of Enviro Energies

Figure 30. Courtesy of David Trumper of MIT

Figure 31. Courtesy of Ralph Hollins of Carnegie Mellon University

Figure 32. Courtesy of David Trumper

Figure 33. Courtesy of U.S. Air Force and General Atomics

Figure 34. Courtesy of National Archives

Figure 35. Courtesy of Phyllis Wilkins, Executive Director of Maglev Maryland, via John Harding

Figure 36. Courtesy of Transrapid International GmbH & Co. Photo by Fritz Stoiber

Figure 37. (top) Photo courtesy of the Railway Technical Research Institute, Shinjuku Office, Tokyo

Figure 37. (bottom) Photo courtesy of Transrapid International, Munich.

Figure 38. Courtesy of Transrapid International GmbH & Co. Photo by Fritz Stoiber

Index

Ingram Content Group UK Ltd.
Milton Keynes UK
UKHW011837080323
418201UK00003B/89/J